S0-BAK-224

Geography

Twenty-first Edition

06/07

EDITOR

Gerald R. Pitzl

Macalester College (retired)

Gerald R. Pitzl received his bachelor's degree in secondary social science education from the University of Minnesota in 1964 and his M.A. (1971) and Ph.D. (1974) in geography from the same institution. Dr. Pitzl has taught a wide array of geography courses, and he is the author of a number of articles on geography, the developing world, and the use of the Harvard Case Method. His book, *Encyclopedia of Human Geography,* was published in January 2004. This one-volume work is designed for use in the Advanced Placement (AP) Program in Human Geography. Dr. Pitzl continues to conduct workshops and in-service sessions on active learning and case-based discussion methods. Since 2002, he has served as an educational consultant with the New Mexico Public Education Department in Santa Fe. His most recent assignment is with the Rural Education Division.

Contemporary Learning Series

2460 Kerper Blvd., Dubuque, IA 52001

Visit us on the Internet
http://www.mhcls.com

Credits

1. **Geography in a Changing World**
   Unit photo—© Getty Images/PhotoLink/Scenics of America
2. **Human-Environment Relationships**
   Unit photo—© Getty Images/PhotoLink/C. Sherburne
3. **The Region**
   Unit photo—© Getty Images/Photodisc Collection
4. **Spatial Interactions and Mapping**
   Unit photo— © Getty Images/Photodisc Collection
5. **Population, Resources, and Socioeconomic Development**
   Unit photo—© Getty Images/Punchstock

Copyright

Cataloging in Publication Data
Main entry under title: Annual Editions: Geography. 2006/2007.
1. Geography—Periodicals. I. Pitzl, Gerald R., *comp.* II. Title: Geography.
ISBN 0–07–354567-8   658'.05   ISSN 1091–9937

Twenty-first Edition

Cover image ©
Printed in the United States of America   1234567890QPDQPD98765   Printed on Recycled Paper

Editors/Advisory Board

Members of the Advisory Board are instrumental in the final selection of articles for each edition of ANNUAL EDITIONS. Their review of articles for content, level, currentness, and appropriateness provides critical direction to the editor and staff. We think that you will find their careful consideration well reflected in this volume.

Preface

In publishing ANNUAL EDITIONS we recognize the enormous role played by the magazines, newspapers, and journals of the public press in providing current, first-rate educational information in a broad spectrum of interest areas. Many of these articles are appropriate for students, researchers, and professionals seeking accurate, current material to help bridge the gap between principles and theories and the real world. These articles, however, become more useful for study when those of lasting value are carefully collected, organized, indexed, and reproduced in a low-cost format, which provides easy and permanent access when the material is needed. That is the role played by ANNUAL EDITIONS.

The articles in this twenty-first edition of *Annual Editions: Geography* represent the wide range of topics associated with the discipline of geography. The major themes of spatial relationships, regional development, the population explosion, and socioeconomic inequalities exemplify the diversity of research areas within geography.

The book is organized into five units, each containing articles relating to geographical themes. Selections address the conceptual nature of geography and the global and regional problems in the world today. This latter theme reflects the geographer's concern with finding solutions to these serious issues. Regional problems, such as food shortages in the Sahel and the greenhouse effect, concern not only geographers but also researchers from other disciplines.

The association of geography with other fields is important, because expertise from related research will be necessary in finding solutions to some difficult problems. Input from the focus of geography is vital in our common search for solutions. This discipline has always been integrative. That is, geography uses evidence from many sources to answer the basic questions, "Where is it?" "Why is it there?" and "What is its relevance?" The first group of articles emphasizes the interconnectedness not only of places and regions in the world but of efforts toward solutions to problems as well. No single discipline can have all of the answers to the problems facing us today; the complexity of the issues is simply too great.

The writings in unit 1 discuss particular aspects of geography as a discipline and provide examples of the topics presented in the remaining four sections. Units 2, 3, and 4 represent major themes in geography. Unit 5 addresses important problems faced by geographers and others.

Annual Editions: Geography 06/07 will be useful to both faculty members and students in their study of geography. The anthology is designed to provide detail and case study material to supplement the standard textbook treatment of geography. The goals of this anthology are to introduce students to the richness and diversity of topics relating to places and regions on the Earth's surface, to pay heed to the serious problems facing humankind, and to stimulate the search for more information on topics of interest. As such, this anthology is an ideal companion volume for use in the Advanced Placement (AP) Program in Human Geography found in high schools across the country. This program, like others sponsored by The College Board, has grown steadily since its inception. Currently, over 10,000 high school students are enrolled in the AP Human Geography Program.

I would like to express my gratitude to Devi Benjamin for her encouragement and invaluable assistance in preparing the twentieth edition. Without her enthusiasm and professional efforts, this project would not have moved along as efficiently as it did. Special thanks are also extended to the McGraw-Hill Contemporary Learning Series editorial staff for coordinating the production of the reader. A word of thanks must go as well to all those who recommended articles for inclusion in this volume and who commented on its overall organization. Especially helpful this year was Peter O. Muller, Thomas R. Paradise, and Leon Yacher.

In order to improve the next edition of *Annual Editions: Geography,* we need your help. Please share your opinions by filling out and returning to us the postage-paid Article Rating Form found on the last page of this book. We will give serious consideration to all your comments and recommendations.

Gerald R. Pitzl
Editor

Contents

UNIT 1
Geography in a Changing World

The concepts in bold italics are developed in the article. For further expansion, please refer to the Topic Guide and the Index.

UNIT 2
Human-Environment Relations

The concepts in bold italics are developed in the article. For further expansion, please refer to the Topic Guide and the Index.

UNIT 3
The Region

The concepts in bold italics are developed in the article. For further expansion, please refer to the Topic Guide and the Index.

UNIT 4
Spatial Interactions and Mapping

The concepts in bold italics are developed in the article. For further expansion, please refer to the Topic Guide and the Index.

UNIT 5
Population, Resources, and Socioeconomic Development

The concepts in bold italics are developed in the article. For further expansion, please refer to the Topic Guide and the Index.

The concepts in bold italics are developed in the article. For further expansion, please refer to the Topic Guide and the Index.

Topic Guide

This topic guide suggests how the selections in this book relate to the subjects covered in your course. You may want to use the topics listed on these pages to search the Web more easily.

On the following pages a number of Web sites have been gathered specifically for this book. They are arranged to reflect the units of this *Annual Edition.* You can link to these sites by going to the student online support site at *http://www.mhcls.com/online/.*

ALL THE ARTICLES THAT RELATE TO EACH TOPIC ARE LISTED BELOW THE BOLD-FACED TERM.

Accessibility
27. Calling All Nations
31. A City of 2 Million Without a Map

Agriculture
9. Global Warming
11. A Great Wall of Waste
18. Central Washington's Emerging Hispanic Landscape
19. Drying Up
35. Farms Destroyed, Stricken Sudan Faces Food Crisis
37. Turning Oceans Into Tap Water
38. Putting the World to Rights
39. Mexico: Was NAFTA Worth It?

AIDS
7. After Apartheid
32. AIDS Infects Education System in Africa
34. China's Secret Plague
38. Putting the World to Rights

Apartheid
7. After Apartheid

Development
13. The Rise of India
38. Putting the World to Rights
39. Mexico: Was NAFTA Worth It?

Drought
9. Global Warming
18. Central Washington's Emerging Hispanic Landscape
19. Drying Up
36. Dry Spell

Economic growth
7. After Apartheid
8. The Race to Save a Rainforest
11. A Great Wall of Waste
15. A Dragon with Core Values
16. Where Business Meets Geopolitics
18. Central Washington's Emerging Hispanic Landscape
19. Drying Up
22. An Inner-City Renaissance
37. Turning Oceans Into Tap Water

Economic issues
10. Environmental Enemy No. 1

Environment
10. Environmental Enemy No. 1
25. Internet GIS: Power to the People!

Geographic Information Systems (GIS)
23. Mapping Opportunities
24. Geospatial Asset Management Solutions
25. Internet GIS: Power to the People!
26. The Future of Imagery and GIS

Geography
1. The Big Questions in Geography
2. Rediscovering the Importance of Geography
4. The Power of Place
5. The Changing Landscape of Fear
23. Mapping Opportunities

Geography, history of
3. The Four Traditions of Geography

Geopolitics
5. The Changing Landscape of Fear
14. Between the Mountains
16. Where Business Meets Geopolitics
29. Fortune Teller

Global issues
5. The Changing Landscape of Fear
6. Watching Over the World's Oceans
8. The Race to Save a Rainforest
9. Global Warming
10. Environmental Enemy No. 1
13. The Rise of India
16. Where Business Meets Geopolitics
21. Deep Blue Thoughts
23. Mapping Opportunities
27. Calling All Nations
33. The Longest Journey
37. Turning Oceans Into Tap Water

Kyoto protocol
38. Putting the World to Rights

Landscape
5. The Changing Landscape of Fear

Maps
25. Internet GIS: Power to the People!
27. Calling All Nations
28. Mapping the Nature of Diversity
29. Fortune Teller
31. A City of 2 Million Without a Map

Migration
33. The Longest Journey
39. Mexico: Was NAFTA Worth It?

Oceans
6. Watching Over the World's Oceans
21. Deep Blue Thoughts

Place
4. The Power of Place

Pollution
9. Global Warming
11. A Great Wall of Waste

Internet References

The following internet sites have been carefully researched and selected to support the articles found in this reader. The easiest way to access these selected sites is to go to our student online support site at *http://www.mhcls.com/online/*.

AE: Geography 06/07

The following sites were available at the time of publication. Visit our Web site—we update our student online support site regularly to reflect any changes.

General Sources

About: Geography
http://geography.about.com

This Web site, created by the About network, contains hyperlinks to many specific areas of geography, including cartography, population, country facts, historic maps, physical geography, topographic maps, and many others.

The Association of American Geographers (AAG)
http://www.aag.org

Surf this site of the Association of American Geographers to learn about AAG projects and publications, careers in geography, and information about related organizations.

Geography Network
http://www.geographynetwork.com

The Geography Network is an online resource to discover and access geographical content, including live maps and data, from many of the world's leading providers.

National Geographic Society
http://www.nationalgeographic.com

This site provides links to National Geographic's huge archive of maps, articles, and other documents. Search the site for information about worldwide expeditions of interest to geographers.

The New York Times
http://www.nytimes.com

Browsing through the archives of the *New York Times* will provide you with a wide array of articles and information related to the different subfields of geography.

Social Science Internet Resources
http://www.wcsu.ctstateu.edu/library/ss_geography_cartography.html

This site is a definitive source for geography-related links to universities, browsers, cartography, associations, and discussion groups.

U.S. Geological Survey (USGS)
http://www.usgs.gov

This site and its many links are replete with information and resources for geographers, from explanations of El Niño, to mapping, to geography education, to water resources. No geographer's resource list would be complete without frequent mention of the USGS.

UNIT 1: Geography in a Changing World

Alternative Energy Institute (AEI)
http://www.altenergy.org

The AEI will continue to monitor the transition from today's energy forms to the future in a "surprising journey of twists and turns." This site is the beginning of an incredible journey.

Mission to Planet Earth
http://science.hq.nasa.gov/

This site will direct you to information about NASA's Mission to Planet Earth program and its Science of the Earth System. Surf here to learn about satellites, El Niño, and even "strategic visions" of interest to geographers.

Poverty Mapping
http://www.povertymap.net

Poverty maps can quickly provide information on the spatial distribution of poverty. Here you will find maps, graphics, data, publications, news, and links that provide the public with poverty mapping from the global to the subnational level.

Solstice: Documents and Databases
http://solstice.crest.org

In this online source for sustainable energy information, the Center for Renewable Energy and Sustainable Technology (CREST) offers information and databases on renewable energy, energy efficiency, and sustainable living. The site also offers related Web sites, case studies, and policy issues. Solstice also connects to CREST's Web presence.

UNIT 2: Human-Environment Relations

Alliance for Global Sustainability (AGS)
http://www.global-sustainability.org

The AGS is a cooperative venture seeking solutions to today's urgent and complex environmental problems. Research teams from four universities study large-scale, multidisciplinary environmental problems that are faced by the world's ecosystems, economies, and societies.

Human Geography
http://www.geog.le.ac.uk/cti/hum.html

The CTI Centre for Geography, Geology, and Meteorology provides this site, which contains links to human geography in relation to agriculture, anthropology, archaeology, development geography, economic geography, geography of gender, and many others.

The North-South Institute
http://www.nsi-ins.ca

Searching this site of the North-South Institute—which works to strengthen international development cooperation and enhance gender and social equity—will help you find information on a variety of development issues.

www.mhcls.com/online/

United Nations Environment Programme (UNEP)
http://www.unep.ch
Consult this home page of UNEP for links to critical topics of concern to geographers, including desertification and the impact of trade on the environment. The site will direct you to useful databases and global resource information.

US Global Change Research Program
http://www.usgcrp.gov
This government program supports research on the interactions of natural and human-induced changes in the global environment and their implications for study. Find details on the atmosphere, climate change, global carbon and water cycles, ecosystems, and land use plus human contributions and responses.

World Health Organization
http://www.who.int
This home page of the World Health Organization will provide you with links to a wealth of statistical and analytical information about health in the developing world.

UNIT 3: The Region

AS at UVA Yellow Pages: Regional Studies
http://xroads.virginia.edu/~YP/regional.html
Those interested in American regional studies will find this site a gold mine. Links to periodicals and other informational resources about the Midwest/Central, Northeast, South, and West regions are provided here.

Can Cities Save the Future?
http://www.huduser.org/publications/econdev/habitat/prep2.html
This press release about the second session of the Preparatory Committee for Habitat II is an excellent discussion of the question of global urbanization.

NewsPage
http://www.individual.com
Individual, Inc., maintains this business-oriented Web site. Geographers will find links to much valuable information about such fields as energy, environmental services, media and communications, and health care.

World Regions & Nation States
http://www.worldcapitalforum.com/worregstat.html
This site provides strategic and competitive intelligence on regions and individual states, geopolitical analyses, geopolitical factors of globalization, geopolitics of production, and much more.

UNIT 4: Spatial Interactions and Mapping

Edinburgh Geographical Information Systems
http://www.geo.ed.ac.uk/home/gishome.html
This valuable site, hosted by the Department of Geography at the University of Edinburgh, provides information on all aspects of Geographic Information Systems and provides links to other servers worldwide. A GIS reference database as well as a major GIS bibliography is included.

Geography for GIS
http://www.ncgia.ucsb.edu/cctp/units/geog_for_GIS/GC_index.html
This hyperlinked table of contents was created by Robert Slobodian of Malaspina University. Here you will find information regarding GIS technology.

GIS Frequently Asked Questions and General Information
http://www.census.gov/geo/www/faq-index.html
Browse through this site to get answers to FAQs about Geographic Information Systems. It can direct you to general information about GIS as well as guidelines on such specific questions as how to order U.S. Geological Survey maps. Other sources of information are also noted.

International Map Trade Association
http://www.maptrade.org
The International Map Trade Association offers this site for those interested in information on maps, geography, and mapping technology. Lists of map retailers and publishers as well as upcoming IMTA conferences and trade shows are noted.

PSC Publications
http://www.psc.isr.umich.edu
Use this site and its links from the Population Studies Center of the University of Michigan for spatial patterns of immigration and discussion of white and black flight from high immigration metropolitan areas in the United States.

UNIT 5: Population, Resources, and Socioeconomic Development

African Studies WWW (U.Penn)
http://www.sas.upenn.edu/African_Studies/AS.html
Access to rich and varied resources that cover such topics as demographics, migration, family planning, and health and nutrition is available at this site.

Geography and Socioeconomic Development
http://www.ksg.harvard.edu/cid/andes/Documents/Background%20Papers/Geography&Socioeconomic%20Development.pdf
John L. Gallup wrote this 19-page background paper examining the state of the Andean region. He explains the strong and pervasive effects geography has on economic and social development.

Human Rights and Humanitarian Assistance
http://www.etown.edu/vl/humrts.html
Through this site, part of the World Wide Web Virtual Library, you can conduct research into a number of human-rights topics in order to gain a greater understanding of the issues affecting indigenous peoples in the modern era.

Hypertext and Ethnography
http://www.umanitoba.ca/faculties/arts/anthropology/tutor/aaa_presentation.new.html
This site, presented by Brian Schwimmer of the University of Manitoba, will be of great value to people who are interested in culture and communication. He addresses such topics as multivocality and complex symbolization, among many others.

Research and Reference (Library of Congress)
http://lcweb.loc.gov/rr/
This research and reference site of the Library of Congress will lead you to invaluable information on different countries. It provides links to numerous publications, bibliographies, and guides in area studies that can be of great help to geographers.

www.mhcls.com/online/

Space Research Institute

http://arc.iki.rssi.ru/eng/

Browse through this home page of Russia's Space Research Institute for information on its Environment Monitoring Information Systems, the IKI Satellite Situation Center, and its Data Archive.

World Population and Demographic Data

http://geography.about.com/cs/worldpopulation/

On this site you will find information about world population and additional demographic data for all the countries of the world.

We highly recommend that you review our Web site for expanded information and our other product lines. We are continually updating and adding links to our Web site in order to offer you the most usable and useful information that will support and expand the value of your Annual Editions. You can reach us at: *http://www.mhcls.com/annualeditions/*.

World Map

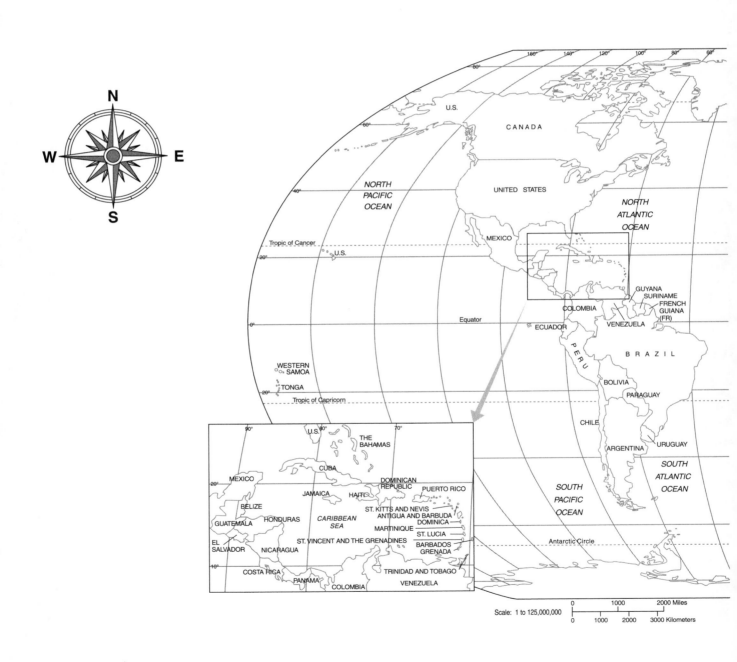

N
W   E
S

Scale: 1 to 125,000,000

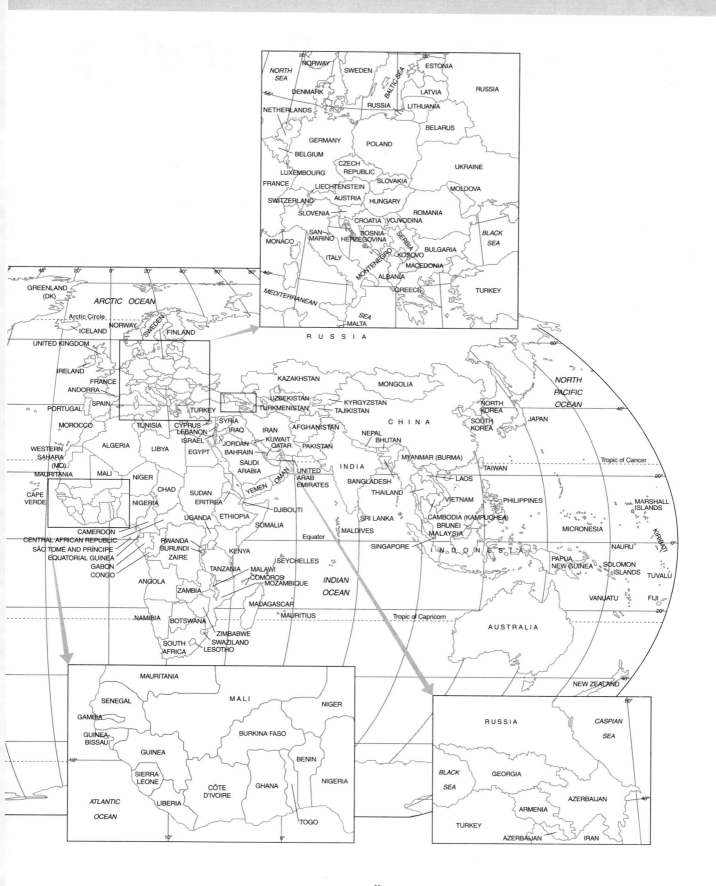

UNIT 1
Geography in a Changing World

Unit Selections

1. **The Big Questions in Geography**, Susan L. Cutter, Reginald Golledge, and William L. Graf
2. **Rediscovering the Importance of Geography**, Alexander B. Murphy
3. **The Four Traditions of Geography**, William D. Pattison
4. **The Power of Place**, Sharon Bishop
5. **The Changing Landscape of Fear**, Susan L. Cutter, Douglas B. Richardson, and Thomas J. Wilbanks
6. **Watching Over the World's Oceans**, Keith Alverson
7. **After Apartheid**, Judith Fein

Key Points to Consider

- Why is geography called an integrating discipline?

- How is geography related to earth science? Give some examples of these relationships.

- What are area studies? Why is the spatial concept so important in geography? What is your definition of geography?

- How has GIS changed geography and cartography?

- What does interconnectedness mean in terms of places? Give examples of how you as an individual interact with people in other places. How are you "connected" to the rest of the world?

- How would you describe your personal "sense of place?"

- In what ways are the oceans crucial to human well-being?

Student Website
www.mhcls.com/online

Internet References
Further information regarding these websites may be found in this book's preface or online.

Alternative Energy Institute (AEI)
  http://www.altenergy.org
Mission to Planet Earth
  http://science.hq.nasa.gov/
Poverty Mapping
  http://www.povertymap.net
Solstice: Documents and Databases
  http://solstice.crest.org

What is geography? This question has been asked innumerable times, but it has not elicited a universally accepted answer, even from those who are considered to be members of the geography profession. The reason lies in the very nature of geography as it has evolved through time. Geography is an extremely wide-ranging discipline, one that examines appropriate sets of events or circumstances occurring at specific places. Its goal is to answer certain basic questions.

The first question- Where is it?-establishes the location of the subject under investigation. The concept of location is very important in geography, and its meaning extends beyond the common notion of a specific address or the determination of the latitude and longitude of a place. Geographers are more concerned with the relative location of a place and how that place interacts with other places both far and near. Spatial interaction and the determination of the connections between places are important themes in geography.

Once a place is "located," in the geographer's sense of the word, the next question is, Why is it here? For example, why are people concentrated in high numbers on the North China plain, in the Ganges River Valley in India, and along the eastern seaboard in the United States? Conversely, why are there so few people in the Amazon basin and the Central Siberian lowlands? Generally, the geographer wants to find out why particular distribution patterns occur and why these patterns change over time.

The element of time is another extremely important ingredient in the geographical mix. Geography is most concerned with the activities of human beings, and human beings bring about change. As changes occur, new adjustments and modifications are made in the distribution patterns previously established. Patterns change, for instance, as new technology brings about new forms of communication and transportation and as once-desirable locations decline in favor of new ones. For example, people migrate from once-productive regions such as the Sahel when a disaster such as drought visits the land. Geography, then, is greatly concerned with discovering the underlying processes that can explain the transformation of distribution patterns and interaction forms over time. Geography itself is dynamic, adjusting as a discipline to handle new situations in a changing world.

Geography is truly an integrating discipline. The geographer assembles evidence from many sources in order to explain a particular pattern or ongoing process of change. Some of this evidence may even be in the form of concepts or theories borrowed from other disciplines. The first article in this unit discusses what the authors consider to be the big questions in geography. The second considers geography's renaissance in U.S. education and its "rediscovery."

Throughout its history, four main themes have been the focus of research work in geography. These themes or traditions, according to William Pattison in "The Four Traditions of Geography," link geography with earth science, establish it as a field that studies land/human relationships, engage it in area studies, and give it a spatial focus. Although Pattison's article first appeared over 30 years ago, it is still referred to and cited frequently today. Much of the geographical research and analysis engaged in today would fall within one or more of Pattison's traditional areas, but new areas are also opening up for geographers.

"The Power of Place" brings the important geographical concept of place into the American high school. The next selection was chosen from the highly acclaimed book, *The Geographical Dimensions of Terrorism*. The introductory chapter, "The Changing Landscape of Fear," states that geography can play a key role in both understanding and combating terrorism. The next article proposes greater attention to the oceans following the devastating tsunami of December 2004.

The Big Questions in Geography

In noting his fondness for geography, John Noble Wilford, science correspondent for *The New York Times*, nevertheless challenged the discipline to articulate those big questions in our field, ones that would generate public interest, media attention, and the respect of policymakers. This article presents our collective judgments on those significant issues that warrant disciplinary research. We phrase these as a series of ten questions in the hopes of stimulating a dialogue and collective research agenda for the future and the next generation of geographic professionals.

Susan L. Cutter
University of South Carolina

Reginald Golledge
University of California, Santa Barbara

William L. Graf
University of South Carolina

Introduction

At the 2001 national meeting of the Association of American Geographers (AAG) in New York City, the opening session featured an address by John Noble Wilford, science correspondent for *The New York Times*. In very candid language, Wilford challenged the discipline to articulate the big questions in our field—questions that would capture the attention of the public, the media, and policymakers (Abler 2001). The major questions posed by Wilford's remarks include the following: Are geographers missing big questions in their research? Why is the research by geographers on big issues not being reported? And what role can the AAG play in improving geographic contributions to address big issues?

First, geographers are doing research on some major issues facing modern society, but not all of them. Geographic thinking is a primary component of the investigation of global warming, for example. Products of that research seen by decision makers and the public often take the form of maps and remote sensing images that explain the geographic outcomes of climate change. Geographic approaches are at the heart of much of the analysis addressing natural and technological hazards, with public interaction taking place through the mapping media. Earthquake, volcanic, coastal, and riverine hazards are all subject to spatial analysis that has become familiar to the public. The terrorist attacks of 11 September 2001 have stimulated new interest in geographic information systems that can be used in response to hazardous events and as guidance in emergency preparedness and response (Figure 1).

In addition to these recent challenges, however, there are major issues that geographers are not addressing adequately at the present time, as illustrated by the accounting that follows in this article. A primary reason for the disconnect between capability to help solve problems and the application of those skills for many major issues is the sociology of the discipline of geography. The majority of AAG members, for example, are academicians, and their agendas and reward structures are targeted at specialized research deeply buried in paradigms that are obscure to decision makers and the public. Additionally, this social structure tends to lead geographic researchers into investigations on small problems that can be solved quickly, produce professional publications, and support a drive for promotion and tenure, rather than investigating more complex, bigger problems that are not easily or quickly solved and do not necessarily lead to academic publications of a type the genre usually demands.

With few exceptions, those geographers outside the university setting are scattered and work individually, in small groups, or as members of larger interdisciplinary teams for governmental agencies, businesses, or private organizations. Because there are few true "institutes" of geographic research, it is difficult to focus geographic energy on big problems. Many geographers in these settings are responding to immediate and short-term demands on their time and talents, rather than leading the larger-scale investigations.

The work in which many geographers engage to address major problems is not reported for two reasons: it does not fit the classic mold for the research journals where geographers get their greatest career awards, and work related to policy often emerges without attribution to the researchers of origin. A significant example that illustrates this point is the work of the Committee on Geography of the Board on Earth Sciences and Resources in the National Research Council (NRC). The committee oversees study committees, which produce geographic studies and reports to guide the federal government in a wide variety of issues that qualify as big questions. Recent work, contributed to primarily by geographers and accomplished from a geographic perspective, includes advice to the U.S. Geological Survey on

Figure 1 *Manhattan, New York, before the terrorist attacks of 11 September 2001 (left), and after (right). Photos by S. L. Cutter.*

reformulating its research programs to address geographic issues entangled in urban expansion, hazards, and mapping. In other cases, one study committee is producing direction for the federal government on what decision makers and the general public need to know about the world of Islam, while another is investigating transportation issues related to urban congestion and the development of livability indicators. Other geographers participate in the Water Science and Technology Board of the NRC, with recent contributions including the use of the watershed concepts in ecosystem management and the role of dams in the security of public water supplies. Another example involves a geographer-led multidisciplinary group to investigate spatial thinking, and another geographer led a major effort in global mapping. In all of these cases, geographers play a central role, but the product of their work is ascribed only to an organization (the NRC), and individuals are recognized only in lists of contributors. If the reports successfully influence policy, the decision makers who actuate that policy take credit for the process, rather than the original investigators who made the recommendations.

The AAG plays a role in stimulating the research that addresses big questions of importance to modern society by recognizing such work and publicizing it. It may be that individual researchers will be more willing to undertake such research if their work is recognized by their colleagues in the discipline as being important and worthy of praise. The AAG can influence the National Science Foundation, the National Endowment for the Humanities, the National Institutes of Health, the National Geographic Society, and other funding sources to channel attention and resources to individuals or teams examining the big questions. Individual geographers are not likely to be able to exert much influence, except when they serve on review panels for these organizations, but the AAG can exert its influence from its steady and visible presence in Washington.

In trying to identify those issues that might qualify as big questions (Table 1), we have included wide-ranging concepts that encompass some conceptual issues (such as scale), but also point out specific topical areas that seem to demand particular attention at the moment. We argue that such diverse big ideas belong in this accounting because, in the end, they are related to each other and mutually supportive. Some of these big questions may be obscure to the public, but most of them are familiar to researchers and policymakers alike, who have already begun to address them. There is little hope that any collection of big questions can identify problems of equal "bigness," but the ones we have identified all seem to warrant teams of researchers and significant funding rather than following the discipline's usual mode of a single or small group of investigators with funding limited to one or two years in duration. The communication of geographic research findings to the public in thoughtful, useful ways represents a major challenge. This challenge, by itself, can also be regarded as yet another big problem facing the discipline. With these introductory comments in mind, we now turn to those questions that we feel are important for the geographic community to address.

What Makes Places and Landscapes Different from One Another and Why Is This Important?

This first question goes to the core of the discipline the relevance of similarities and differences among people, places, and regions. What is the nature of uneven economic development and what can geography contribute to understanding this phenomenon? More specifically, how can national and global policies be implemented in a world that is increasingly fragmented politically, socially, culturally, and environmentally?

To elaborate on this question, we accept an assumption that the human mind is not constructed to handle large-scale continuous

3

Table 1 *Big Questions in Geography*

1. What makes places and landscapes different from one another, and why is this important?

2. Is there a deeply held human need to organize space by creating arbitrary borders, boundaries, and districts?

3. How do we delineate space?

4. Why do people, resources, and ideas move?

5. How has the earth been transformed by human action?

6. What role will virtual systems play in learning about the world?

7. How do we measure the unmeasurable?

8. What role has geographical skill played in the evolution of human civilization, and what role can it play in predicting the future?

9. How and why do sustainability and vulnerability change from place to place and over time?

10. What is the nature of spatial thinking, reasoning, and abilities?

chaos. Nor does it function optimally when dealing with large-scale perfect uniformity. Between these two extremes there is variability, which is the dominant characteristic of both the natural world and the human world. To understand the nature of physical and human existence, we need to examine the occurrence and distribution of variability in various domains. For geographers, this examination has involved exploring the nature of spatial distributions, patterns, and associations, examining the effects of scale, and developing modes of representation that best communicate the outcomes of these explorations. In the course of this search for understanding of the essentials of spatial variation, geographers have attempted to comprehend the interaction between physical and human environments, how people adapt to different environments, and how knowledge about human-environment relations can be communicated through appropriate representational media.

Even in the absence of humans, the earth and the phenomena found on this planet are incredibly diverse. Variability is widespread; uniformity is geographically restricted. Determining the nature and occurrence of variability and uniformity are at the heart of the discipline of geography. No other area of inquiry has, as its primary goal, discovering, representing, and explaining the nature of spatial variability in natural and human environments at scales beyond the microscopic and the figural (body space) such as vista, environmental, or gigantic and beyond (Montello 1993). Most geography has been focused on vista, environmental or gigantic scales, but some (e.g., cognitive behavioral) emphasizes figural scale. Finding patterns or trends towards regularity at some definable scale amidst this variability provides the means for generalizing, modeling, and transferring knowledge from one spatial domain to another. Law-like and theoretical statements can be made, and confidence in the relevance of decisions and policies designed to cope with existence can be determined.

Among other things, geographers have repeatedly found, at some scales, spatial regularity in distributions of occurrences that seem random or indeed chaotic at other scales. Sometimes this results from selecting an appropriate scale and format for summarizing and representing information. Examples include using very detailed environmental-scale data to discover the topologic properties of stream networks, or establishing the regular and random components of human settlement patterns in different environments.

Realizing the spatial variability in all phenomena is a part of the naïve understanding of the world. Being able to explain the nature of variability is the academic challenge that drives the discipline of geography. Like other scientists, geographers examine variability in their search for knowledge and understanding of the world we live in, particularly in the human environment relations and interactions that are a necessity for our continued existence.

Is There a Deeply Held Human Need to Organize Space by Creating Arbitrary Borders, Boundaries, and Districts?

Humans, by their very nature, are territorial. As human civilizations grew from hunter gatherers to more sedentary occupations, physical manifestation of the demarcation of space ensued. Hadrian's Wall kept the Scots and Picts out, the Great Wall of China protected the Ming Dynasties from the Mongols, and the early walled cities of Europe protected those places from barbarians and other acquisitive sociocultural groups.

At a more limited scale, internal spaces in cities were also divided, often based on occupation and/or class. As civilizations grew, space was organized and reorganized into districts that supported certain economic activities. City-states begat nation-states, and eventually most of the world was carved up into political spaces. Nation-states required borders and boundaries (all involving geography), as land and the oceans (and the resources contained within) were carved up into non-equal units. Within nations, land partitioning has been a factor in the decline of environmental quality. For example, the erection of barbed-wire fencing on the Great Plains to separate farming and ranching homesteads from each other did more to hasten the decline of indigenous species and landscape degradation than any other invention at the time (Worster 1979, 1993).

The modern equivalent of the human need and desire for delineating space is the notion of private property. Suburban homes with tall fences between neighbors, for example, help foster the ideal of separation from neighbors and disengagement from the community, both predicated on the need to protect "what's mine" (and of course the ubiquitous property value) as well as providing a basic need for privacy. The tendency for the rich to get richer and the poor to get poorer also applies to the values of these divided properties. The diffusion of democratically controlled, market based economies to much of the globe increases the significance of research that explores why we divide space. Pressing research questions include, for example: Are ghettos bursting with poverty-level inhabitants an inevitable consequence of democratic capitalist societies? Are such societies amenable to concerns for social justice? And how would such concerns influence the patterns and distributions of living activities?

We also lack some of the basic understanding of how the physical delineation of space affects our perception of it. Furthermore, we need better knowledge of how perceptions of physical space alter social, physical, and environmental processes. Finally, has globalization changed our view of the social construction of space? Does physical space still support spatial relations and spatial interactions, or are they becoming somewhat independent, as may be the case in social space, intellectual space, and cyberspace? How will the interactions between people, places, and regions change as our view of space (and time, for that matter) changes?

How Do We Delineate Space?

Once we understand *why* we partition space, we face a closely related issue: *how* do we do it? The definition of regions by drawing boundaries is deceptively simple. The criteria by which we delineate space have far-reaching consequences, because the resulting divisions of space play a large role in determining how we perceive the world. A map of the United States showing the borders of the states, for example, evokes a very different perception of the nation than a similar scale map showing the borders of the major river basins. A further difference in perception is created if the map shows major rivers as networks rather than as basins, and the resulting difference between perception of networks and perception of regions can direct knowledge and its application in divergent ways (National Research Council 1999, x). For example, should we conduct pollution oriented research on rivers or on watersheds, or on the state administrative units that potentially might control pollution? What are the implications of our choice of geographic framework?

The logical, rational delineation of spaces on the globe depends on the criteria to be used, but geographic research offers few established, widely accepted rules about what these criteria should be or how they might be employed. The designation of political boundaries without respect to ethnic cultures has wrought havoc in much of post colonial Africa and central Europe, for example, but geographers have not yet offered workable alternatives that account for the complexities of multicultural populations. In natural-science research and management, a major issue is the establishment of meaningful regions that can be aggregated together to scale up, or that can be disaggregated to scale down. Natural scientists also experience significant difficulty in designing compatible regions across topical subjects. For example, the blending of watersheds, ecosystems, and ranges of particular species poses significant problems in environmental management. Adding to the complexity from a management and policy perspective is the tangle of administrative regions, whose boundaries are often derived from political boundaries rather than natural ones. Recognition of these problems is easy but offering thoughtful geographic solutions to them is not.

Geographers have much to contribute to the delineation of space by developing new knowledge and techniques for defining subdivisions of earth space based on specific criteria, including economic efficiency, compatibility across applications, ease of aggregation and disaggregation, repeatability, and universality of application. Geographers need to develop methods for delineating space that either resist change over time or accommodate temporal changes smoothly.

A continuing example of delineating space that has important political implications is the process of defining American congressional districts once each decade based on the population census. The need for fair representation, relative uniformity in population numbers in each district, recognition of traditional communities, and accommodation of changing population distributions comprise some of the criteria that need not equate to partisan politics in constructing at least the first approximation of redrawn district boundaries (Monmonier 2001). Some states have nonpartisan commissions to delineate the districts, yet geography provides very little substantive advice on the subject to guide such groups.

Why do People, Resources, and Ideas Move?

One of the fundamental concepts in geography is the understanding that goods, services, people, energy, materials, money, and even ideas flow through networks and across space from place to place. Although geography faces questions about all these movements, one of the most pressing questions concerns the movement of people. We have some knowledge about the behavior of people who move their residences from one place to another, and we can observe obvious economic forces leading to the migration of people toward locations of relative economic prosperity. However, we have much less understanding about the episodic movements of people in cities. In most developed countries, the congestion of vehicular traffic has become a significant negative feature in assessing the quality of life, and in lesser-developed countries the increasing number of vehicles used in the context of inadequate road networks results in frustrating delays. Geography can and should address fundamental issues such as the environmental consequences of the decision to undertake laborious journeys to work (e.g., contributions of vehicle exhaust to air pollution, the possible environmental changes induced by telecommuting, and the need for alternative-fuel, low-pollution vehicles). In addition to understanding the environmental consequences of daily moves, the discipline has much to offer in describing, explaining, and predicting the sociocultural consequences of these decisions.

The flow of vehicles on roadways involves obvious physical networks, but there are other flows demanding attention that operate through more abstract spaces. The diffusion of culture—particularly "Western" culture, with its emphasis on materialism and individualism—is one of the leading edges of globalization of the world economy. Geographers must begin to address how these social, cultural, and economic forces operate together to diffuse, from a few limited sources, an extensive array of ideas and attitudes that are accepted by a diverse set of receiving populations. Even if such diffusion takes place through digital space, it probably does so in a distinctive geography that we should understand if we are to explain and predict the world in the twenty-first century.

The electrical energy crisis of 2001 made us aware, quite vividly, of the finiteness of nonrenewable resources such as oil and gas and of the difficulties in their distributions. We have already consumed more than 50 percent of the world's known reserves of these resources. Historically, as one energy source has replaced another (as when coal power replaced water power), there have been changes in the locational patterns, growth, importance of settlements, and significance of regions. Examples include the decline of heavy industrial areas into "rust belts" and their replacement with service- and information-based centers that have more locational flexibility. As current energy sources change, what will happen to urban location and growth? Will the geopolitical power structure of the world change markedly? For example, will the countries that are part of the Organization of the Petroleum Exporting Countries (OPEC) retain their global economic power and political strength? Will existing populations and settlements decline, or relocate to alternative sources of energy? What will be the geographic configuration of the economic and political power that goes with such changes?

Finally, the more physically oriented flows, such as those of energy and materials, present a demanding set of questions for geographers. While geochemists are deriving the magnitudes of elemental fluxes of such substances as carbon and nitrogen, for example, it is incumbent on geographers to point out that these fluxes do not take place in aspatial abstract ways, but rather in a physically and socially defined landscape that has important locational characteristics. In other words, although there may very well be an understanding of the amounts of nitrogen circulating from earth to oceans to atmosphere, that circulation is not everywhere equal. How does human management affect the nitrogen and other elemental cycles? What explains its geographic vari-

Figure 2 *A local example of transformations brought about in the natural world by humans. The lower Sandy River of western Oregon appears to be a pristine river, but it has radically altered water, sediment, and biological systems because of upstream dams. Photo by W. L. Graf.*

ability? How does that variability change in response to controls not related to human intervention? This leads to our next big question.

How Has the Earth Been Transformed by Human Action?

Humans have altered the earth, its atmosphere, and its water on scales ranging from local to global (Thomas 1956; Turner et al. 1990). At the local scale, many cities and agricultural landscapes represent nearly complete artificiality in a drive to create comfortable places in which to live and work, and to maximize agricultural production for human benefit. The transformations have also had negative effects at local scales, such as altering the chemical characteristics of air and water, converting them into media that are toxic for humans as well as other species. At regional scales, human activities have resulted in wholesale changes in ecosystems, such as the deforestation of northwest Europe over the past several centuries, a process that seems to be being replicated in many tropical regions today. At a global scale, the introduction of industrial gases into the atmosphere plays a still emerging role in global climate change. Taken together, these transformations have had a geographically variable effect that geographers must better define and explain. Dilsaver and Colten (1992, 9) succinctly outlined the basic questions almost a decade ago: How have human pursuits transformed the environment, and how have human social organizations exerted their control over environments? Graf (2001) recently asked how we can undo some of our previous efforts at environmental change and control.

In many instances, this explanation of variation might emphasize the physical aspects of changes, or understanding the underlying dynamics of why the changes occur (Dilsaver, Wyckoff, and Preston 2000). Wide-ranging assessments of river basins, for example, must rely on a plethora of controlling factors ranging from land use to water, sediment, and contaminant movements. Geographers must employ more complicated and insightful approaches, however, to truly understand why transformations vary from place to place, largely in response to the connection between the biophysical environment and the human society that occupies it. Understanding this

delicate interplay between nature and society as a two-way connection can lead us to new knowledge about social and environmental landscapes, but it can also help us make better decisions on how to achieve future landscapes that are more often transformed in nondestructive ways.

One of the primary issues facing many societies in their relationship with their supporting environments is how much of the biophysical world should be left unchanged, or at least changed to the minimal degree possible. The amount of remaining "natural" landscape in many nations is small—probably less than 5 percent of the total surface—so time is growing short to decide what areas should be set aside and preserved (Figure 2). Not only do these preservation decisions affect land and water surfaces; they also profoundly affect nonhuman species that use the surfaces for habitat. If human experience is enriched by diverse ecosystems, then the decline in biodiversity impoverishes humanity as well. Which areas should be preserved and why? How should preserved areas be linked with one other? How can public and private property productively coexist with nearby preserved areas?

What Role Will Virtual Systems Play in Learning about the World?

Stated another way, what will virtual systems allow us to do in the future that we cannot do now? What new problems can be pursued (Golledge forthcoming)? Providing an answer opens a Pandora's box of questions concerning the geographic impacts of new technologies (Goodchild 2000). What new multimodal interfaces for interpreting visualized onscreen data need to be developed in order to overcome current technological constraints of geographic data visualization? Can we produce a virtual geography? Do we really want to?

One serious problem that deserves immediate attention is the examination of the geographic implications of the development of economies and societies based on information technology. In particular, the sociospatial implications of an increasing division between the digital haves and have-nots demand attention. Pursuing such a problem will require answering questions about the geographic consequences of employment in cyberspace and its implication for human movements such as migration, intraurban mobility, commuting, and activity-space restructuring. The current extensive demand for and use of transportation for business purposes may need to be re-examined. It may be argued that, in the world of business communication, geographic distance is a decreasingly important factor, because both digital and visual interaction can take place at the click of a mouse button without the need for person-to-person confrontation. If this is so, what are the longer-term impacts for living and lifestyles, and how could the inhabitation and use of geographic environments be affected? If this is true, why is it that we see dramatic concentrations of cyber-businesses in a few areas, similar to the locational behavior of pre-digital industries? Are Silicon Valley in California and Route 128 in Massachusetts simply the "rust belts" of the future?

Research has shown that the most effective way of learning about an environment is by directly experiencing it, so that all sensory modalities are activated during that experience (Figure 3) (Gale 1984; Lloyd and Heivly 1987; MacEachren 1992). However, many places are distant or inaccessible to most people. The interior of the Amazon rainforest, the arctic tundra of northern Siberia, Himalayan peaks, the interior of the Sahara desert, Antarctica and the South Pole, the barrios of Rio de Janeiro, and the Bosnian highlands can become much closer to us. Satellite imagery provides detailed digitized imagery of these

Figure 3 *Exploring immersive virtual worlds with equipment developed between 1992 and 2001, showing the original and the miniaturized versions of a GPS-driven auditory virtual environment at the University of California, Santa Barbara. The more cumbersome 1992 version is shown on the left, with the reduced 2001 version on the right. Psychologist Jack Loomis and associates developed the system, demonstrated here by author Reginald Golledge. Photo courtesy of R. Golledge.*

places. A problem awaiting solution is how to use this extensive digital database to build virtual systems that will allow immersive experiences with such environments. Problems of motion-sickness experienced by some people in immersive systems need to be solved; assuming this will be achieved, virtual reality could become the laboratory of the future for experiencing different places and regions around the world.

Discovering how best to deal with problematic futures, on earth or on other planets, is definitely one of the big problems facing current and future geographers. Many land use planning, transportation, and social policies are made on an "if _____ then _____" basis. Because we are unable to change the world experimentally, we need to investigate other ways of observing environmental events and changes. Examples include changing a street for vehicles to a pedestrian mall to explore human movement behavior, or experiencing the action and consequences of snow or mud avalanches in tourist-dependent alpine environments. What more can we learn by building and manipulating virtual environments? In a virtual system, we can raise local pollution levels, accelerate global warming, change sea levels by melting ice caps, or simulate the impacts of strictly enforcing land conversion policies at the rural-urban fringes of large cities. In the face of an increasingly international economy and globalization of environmental issues, there is a need to develop a way to explore possible scenarios before implementing policies theoretically designed to deal with global (or more local) problems.

How Do We Measure the Unmeasurable?

Geography is normally practiced at local to national scales at which we can get a clear sense of the existence or development of patterns and processes. People, landscapes, and resources are not evenly distributed on the earth's surface, so we begin with a palette that is diverse. How can we accommodate such diversity in policies to avoid winners and losers? Economists, for example, assume away all spatial variability in their economic models. What happens to general models when space is introduced? How can we transform from the local to the global and vice versa? The question of scale transformation, especially the calibration of large-scale global circulation models or the development of climate-impact models globally with local or regional applications is a major area in which geography can contribute and is playing a leading role (AAG GCLP Group forthcoming).

We need to develop more compatible databases that have an explicit geographic component, with geocoded data that permit us to scale up and scale down as the need arises. Data collection, archiving, and dissemination all are areas that require our expertise, be it demographic data, environmental data, or land-use data. The large question is, how do we maintain a global information system that goes beyond the petty tyrannies of nation-states (and the need to protect information for "security" reasons), yet protects individuals' right to privacy? The selective use of remote-sensing techniques to monitor environmental conditions has been helpful in understanding the linkages between local activities and global impacts. However, can we use advanced technology to support demographic data collection and analyses and still maintain safeguards on privacy protection (Liverman et al. 1998)? For example, recent Supreme Court decisions have placed important legal protections on the use of thermal infrared sensors in public safety.

Another series of issues involves the aggregation of human behaviors. How can we geographically aggregate data along a set of common dimensions to insure its representation of reality and get around the thorny issues of averaging and the mean-areal-center or modified areal-unit problems? We often use techniques to handle aggregated populations and areas that in fact, depart from reality, creating a type of artificial environment. Unfortunately, public policies all too often are based on these constructed realities, thus further exacerbating the distribution of goods, services, and resources. What new spatial statistical tools do we need to address this concern?

Lastly, in a post-11-September world, how do we measure the geography of fear? Does the restriction of geographic data (presumably for national security reasons) attenuate or amplify fear of the unknown? The discipline requires the open access to information and data about the world and the people who live there. Data access will be one of the key issues for our community to address in the coming years.

What Role Has Geographical Skill Played in the Evolution of Human Civilization, and What Role Can It Play in Predicting the Future?

Is there a necessary geographic base to human history? If so, how can we improve our ability to predict spatial events and events that have spatial consequences that will fundamentally shape the future? Can we develop the geographic equivalent of leading economic indicators?

From the early cradles of civilization in Africa and Asia, humankind gradually colonized the earth. This process of redistributing people in space (migration) was caused by population growth, resource exhaustion, attractive untapped resources, environmental change, environmental hazard, disease, or invasion and succession by other human groups. But what skills and abilities were required to ensure success in relocation movements? Were the movements random or consciously directed? If they were directed, then what skills and/or abilities were required by explorers, leaders, and followers to ensure success? What criteria had to be satisfied before resettlement was possible? What new geographic skills and abilities have been developed throughout human history, and which ones have deteriorated or disappeared? Have geographic skills and abilities been maintained equally in males and females? If not, what developments in the evolution of human civilizations have mediated such losses or changes?

While we know much about human history, we know little about the geographical basis of world history, and we know little of the extent to which the presence or absence of geographic knowledge played a significant part in historical development. For example, would Napoleon's invasion of Russia been more successful had skilled and knowledgeable geographers counseled him on the route chosen and the appropriate season for movement? Historians often tell us that understanding the past is the key to knowing the present and to successfully predicting the future. We cannot fully understand the past if we ignore or diminish the importance of environmental diversity and knowledge about those variations that are the result of spatial and geographic thinking and reasoning. A similar argument can be made for predicting future events and behaviors. What geographic knowledge is likely to be important in prediction? Must we rely on assumptions about uniform environments, population characteristics, tastes and preferences, customs, beliefs, and values? Such a procedure is precarious at best. However, we do not currently know how to incorporate geographic variability into our models, or indeed what variables should be incorporated into predictive models. Achieving such a goal is a necessary part of increasing our very limited predictive capabilities.

How and Why Do Sustainability and Vulnerability Change from Place to Place and over Time?

Historically, geography was an integrative science with a particular focus on regions. It then switched from breadth to depth, with improvements in theory, methods, and techniques. We are now returning to that earlier perspective as we look for common ground in the interactions between human systems and physical systems. Increasing population pressures, the regional depletion or total exhaustion of resources, environmental degradation, and rampant development are processes that affect the sustainability of natural systems and constructed environments. There is a movement toward the integration of many different social and natural science perspectives into a field called sustainability science (Kates et al. 2001). Understanding what constrains and enhances sustainable environments will be an important research theme in the future. How can we maintain and improve the quality of urban environments for general living (social, economic, and environmental conditions)? How long can the processes of urban and suburban growth continue without deleterious and fundamental changes in the landscape and the escalation in costs of environmental restoration? Suburban sprawl is already a major policy issue. What is the long-term impact on human survival of the constant usurpation of agricultural land by the built environment? How long can we continue slash and burn agriculture in many parts of the tropical world? What triggers the environmental insecurity of nations, and how does this lead to armed conflicts and mass migrations of people? How have these processes varied in time and space? What are the greatest threats to the sustainability of human settlements, agriculture, energy use, for example and how can we mitigate or reduce those threats (NRC 2000)?

Nonsustainable environments enhance the effect of risks and hazards and ultimately increase both biophysical and social vulnerability, often resulting in disasters of one kind or another. When societies or ecosystems lack the ability to stop decay or decline and they do not have the adequate means to defend against such changes, there can be potentially catastrophic results. Examples include the environmental degradation of the Aral Sea, the increasing AIDS pandemic, and the human and environmental costs of coastal living (Heinz Center 2000). Vulnerability can be thought of as a continuum of processes, ranging from the initial susceptibility to harm to resilience (the ability to recover) to longer-term adaptations in response to large-scale environmental changes (Cutter, Mitchell, and Scott 2000). These processes manifest themselves at different geographic scales, ranging from the local to the global. What is the threshold when vulnerability ceases to become something we can deal with and becomes something we cannot? At what point does the built environment or ecosystem extend beyond its own ability to recover from natural or social forces?

What Is the Nature of Spatial Thinking, Reasoning, and Abilities?

Geographic knowledge is the product of spatial thinking and reasoning (Golledge 2002b). These processes require the ability to comprehend scale changes; transformations of phenomena, or representations among one, two, and three spatial dimensions. They also require understanding of: the effect of distance, direction, and orientation on developing spatial knowledge; the nature of reference frames for identifying locations, distributions and patterns; the nature of spatial hierarchies; the nature of forms by extrapolating from cross sections; the significance of adjacency and nearest neighbor concepts; the spatial properties of density, distance, and density decay; and the configurations of patterns and shapes in various dimensions and with differing degrees of completeness. It also requires knowing the implications of spatial association and understanding other concepts not yet adequately articulated or understood. What geography currently lacks is an elaboration of the fundamental geographic concepts and skills that are necessary for the production and communication of spatial and geographic information. In the long run, this will be needed before geography can develop a well-articulated knowledge base of a type similar to other human and physical sciences.

Conclusion

In the American Declaration of Independence, Thomas Jefferson wrote that among the most basic of human rights are life, liberty, and the pursuit of happiness. Each of these rights is played out upon a geographic stage, has geographic properties, and operates as a geographical process. Geography, as a field of knowledge and as a perspective on the world, has paid too little attention to these grand ideas, and they are fertile ground for the seeds of new geographic research. How and why does the opportunity for the pursuit of happiness vary from one place to another, and does the very nature of that pursuit change geographically?

In pursuit of answers to the big questions articulated above, we will inevitably need to think about doing research on problems such as:

- What are the spatial constraints on pursuing goals of life, liberty, and the pursuit of happiness?
- What are our future resource needs, and where will we find the new resources that have not, at this stage, been adequately explored?
- When does geography start and finish? Does it matter?
- What are likely to be the major problems in doing the geography of other planets?
- Will cities of the future remain bound to the land surface, or will they move to what we now consider unlikely or exotic locations (under water or floating in space)?

The big questions posed here are not all encompassing. They represent our collective judgments (and biases) on what issues are significant for the discipline, and those that should provide a focus for our considerable intellectual capital. Not everyone will agree with us, nor should they. We view this article as the beginning of a dialogue within the discipline as to what are the probable big questions for the next generation of geographers.

Literature Cited

Abler, R. F. 2001. From the meridian—Wilford's "science writer's view of geography." *AAG Newsletter* 36 (4): 2, 9.

Association of American Geographers (AAG) Global Change in Local Places (GCLP) Research Group. Forthcoming. *Global change and local places: Estimating, understanding, and reducing greenhouse gases*. Cambridge, U.K.: Cambridge University Press.

Cutter, S. L., J. T. Mitchell, and M. S. Scott. 2000. Revealing the vulnerability of people and places: A case study of Georgetown County, South Carolina. *Annals of the Association of American Geographers* 90:713–37.

Dilsaver, L. M., and C. E. Colten, eds. 1992. *The American environment: Interpretations of past geographies*. Lanham, MD: Rowan and Littlefield Publishers.

Dilsaver, L. M., W. Wyckoff, and W. L. Preston. 2000. Fifteen events that have shaped California's human landscape. *The California Geographer* XL: 1–76.

Gale, N. D. 1984. Route learning by children in real and simulated environments. Ph.D. diss., Department of Geography, University of California, Santa Barbara.

Golledge, R. G. 2002. The nature of geographic knowledge. *Annals of the Association of American Geographers* 92 (1): 1–14.

—. Forthcoming. *Spatial cognition and converging technologies*. Paper presented at the Workshop on Converging Technology (NBIC) for Improving Human Performance, sponsored by the National Science Foundation. Washington, D.C. In press.

Goodchild, M. F. 2000. Communicating geographic information in a digital age. *Annals of the Association of American Geographers* 90:344–55.

Graf, W. L. 2001. Dam age control: Restoring the physical integrity of America's rivers. *Annals of the Association of American Geographers* 91:1–27.

Heinz Center. 2000. *The hidden costs of coastal erosion*. Washington, D.C.: The H. John Heinz III Center for Science, Economics and the Environment.

Kates, R. W., W. C. Clark, R. Corell, J. M. Hall, C. C. Jaeger, I. Lowe, J. J. McCarthy, H. J. Schnellnhuber, B. Bolin, N. M. Dickson, S. Faucheux, G. C. Gallopin, A. Grubler, B. Huntley, J. Jager, N. S. Jodha, R. E. Kasperson, A. Mabogunje, P. Matson, H. Mooney, B. Moore III, T. O'Riodan, and U. Svedin. 2001. Sustainability science. *Science* 292:641–42.

Liverman, D., E. F. Moran, R. R. Rindfuss, and P. C. Stern, eds. 1998. *People and pixels: Linking remote sensing and social science*. Washington, D.C.: National Academy Press.

Lloyd, R. E., and C. Heivly. 1987. Systematic distortion in urban cognitive maps. *Annals of the Association of American Geographers* 77:191–207.

MacEachren, A. M. 1992. Application of environmental learning theory to spatial knowledge acquisition from maps. *Annals of the Association of American Geographers* 82 (2): 245–74.

Monmonier, M. S. 2001. *Bushmanders and Bullwinkles: How politicians manipulate electronic maps and census data to win elections*. Chicago: University of Chicago Press.

Montello, D. R. 1993. Scale and multiple psychologies of space. In *Spatial information theory: A theoretical basis for GIS. Lecture notes in computer science 716. Proceedings, European Conference, COSIT '93. Marciana Marina, Elba Island, Italy, September*, ed. A. U. Frank and I. Campari. 312–21. New York: Springer-Verlag.

National Research Council (NRC). 1999. *New strategies for America's watersheds*. Washington, D.C.: National Academy Press.

—. 2000. *Our common journey: A transition toward sustainability*. Washington, D.C.: National Academy Press.

Thomas, W. L., Jr., ed. 1956. *Man's role in changing the face of the earth*. Chicago: The University of Chicago Press.

Turner, B. L. II, W. C. Clark, R. W. Kates, J. F. Richards, J. T. Mathews, and W. Meyer, eds. 1990. *The earth as transformed by human action: Global and regional changes in the biosphere over the past 300 years*. Cambridge, U.K.: University of Cambridge Press.

Worster, D. E. 1979. *Dust bowl: The Southern plains in the 1930s*. Oxford: Oxford University Press.

—. 1993. *The wealth of nature: Environmental history and the ecological imagination*. New York: Oxford University Press.

SUSAN L. CUTTER is Carolina Distinguished Professor, Department of Geography, University of South Carolina, Columbia, SC 29208. E-mail: scutter@sc.edu. She served as president of the Association of American Geographers from 2000-2001, and is a fellow of the American Association for the Advancement of Science (AAAS). Her research interests are vulnerability science, and environmental hazards policy and management.

REGINALD GOLLEDGE is a Professor of Geography at the University of California, Santa Barbara, Santa Barbara, CA 93106. E-mail: golledge@geog.ucsb.edu and served as AAG president from 1999 to 2000. His research interests include various aspects of behavioral geography (spatial cognition, cognitive mapping, spatial thinking), the geography of disability (particularly the blind), and the development of technology (guidance systems and computer interfaces) for blind users.

WILLIAM L. GRAF is Education Foundation University Professor and Professor of Geography at the University of South Carolina, Columbia, SC 29208. E-mail: graf@sc.du. He served as AAG president from 1998–1999, and is a National Associate of the National Academy of Science. His specialties are fluvial geomorphology and policy for public land and water.

From *The Professional Geographer,* Susan L. Cutter, et al, August 2002, pp. 305-317. Copyright © 2002 by Blackwell Publishers, Ltd. Reprinted by permission.

POINT OF VIEW

Rediscovering the Importance of Geography

By Alexander B. Murphy

As AMERICANS STRUGGLE to understand their place in a world characterized by instant global communications, shifting geopolitical relationships, and growing evidence of environmental change, it is not surprising that the venerable discipline of geography is experiencing a renaissance in the United States. More elementary and secondary schools now require courses in geography, and the College Board is adding the subject to its Advanced Placement program. In higher education, students are enrolling in geography courses in unprecedented numbers. Between 1985–86 and 1994–95, the number of bachelor's degrees awarded in geography increased from 3,056 to 4,295. Not coincidentally, more businesses are looking for employees with expertise in geographical analysis, to help them analyze possible new markets or environmental issues.

In light of these developments, institutions of higher education cannot afford simply to ignore geography, as some of them have, or to assume that existing programs are adequate. College administrators should recognize the academic and practical advantages of enhancing their offerings in geography, particularly if they are going to meet the demand for more and better geography instruction in primary and secondary schools. We cannot afford to know so little about the other countries and peoples with which we now interact with such frequency, or about the dramatic environmental changes unfolding around us.

From the 1960s through the 1980s, most academics in the United States considered geography a marginal discipline, although it remained a core subject in most other countries. The familiar academic divide in the United States between the physical sciences, on one hand, and the social sciences and humanities, on the other, left little room for a discipline concerned with how things are organized and relate to one another on the surface of the earth—a concern that necessarily bridges the physical and cultural spheres. Moreover, beginning in the 1960s, the U.S. social-science agenda came to be dominated by pursuit of more-scientific explanations for human phenomena, based on assumptions about global similarities in human institutions, motivations, and actions. Accordingly, regional differences often were seen as idiosyncrasies of declining significance.

Although academic administrators and scholars in other disciplines might have marginalized geography, they could not kill it, for any attempt to make sense of the world must be based on some understanding of the changing human and physical patterns that shape its evolution—be they shifting vegetation zones or expanding economic contacts across international boundaries. Hence, some U.S. colleges and universities continued to teach geography, and the discipline was often in the background of many policy issues—for example, the need to assess the risks associated with foreign investment in various parts of the world.

By the late 1980s, Americans' general ignorance of geography had become too widespread to ignore. Newspapers regularly published reports of surveys demonstrating that many Americans could not identify major countries or oceans on a map. The real problem, of course, was not the inability to answer simple questions that might be asked on *Jeopardy!*; instead, it was what that inability demonstrated about our collective understanding of the globe.

Geography's renaissance in the United States is due to the growing recognition that physical and human processes such as soil erosion and ethnic unrest are inextricably tied to their geographical context. To understand modern Iraq, it is not enough to know who is in power and how the political system functions. We also need to know something about the country's ethnic groups and their settlement patterns, the different physical environments and resources within the country, and its ties to surrounding countries and trading partners.

Those matters are sometimes addressed by practitioners of other disciplines, of course, but they are rarely central to the analysis. Instead, generalizations are often made at the level of the state, and little attention is given to spatial patterns and practices that play out on local levels or across international boundaries. Such pre-occupations help to explain why many scholars were caught off guard by the explosion of ethnic unrest in Eastern Europe following the fall of the Iron Curtain.

Similarly, comprehending the dynamics of El Niño requires more than knowledge of the behavior of ocean and air currents; it is also important to understand how those currents are situated with respect to land masses and how they relate to other climatic patterns, some of which have been altered by the burning of fossil fuels and other human activities. And any attempt to understand the nature and extent of humans' impact on the environment requires consideration of the relationship between human and physical contributions to environmental change. The factories and cars in a city produce smog, but surrounding mountains may trap it, increasing air pollution significantly.

TODAY, academics in fields including history, economics, and conservation biology are turning to geographers for help with some of their concerns. Paul Krugman, a noted economist at the Massachusetts Institute of Technology, for example, has turned conventional wisdom on its head by pointing out the role of historically rooted regional inequities in how international trade is structured.

Geographers work on issues ranging from climate change to ethnic conflict to urban sprawl. What unites their work is its focus on the shifting organization and character of the earth's surface. Geographers examine changing patterns of vegetation to study global warming; they analyze where ethnic groups live in Bosnia to help understand the pros and cons of competing administrative solutions to the civil war there; they map AIDS cases in Africa to learn how to reduce the spread of the disease.

Geography is reclaiming attention because it addresses such questions in their relevant spatial and environmental contexts. A growing number of scholars in other disciplines are realizing that it is a mistake to treat all places as if they were essentially the same (think of the assumptions in most economic models), or to undertake research on the environment that does not include consideration of the rela-tionships between human and physical processes in particular regions.

Still, the challenges to the discipline are great. Only a small number of primary- and secondary-school teachers have enough training in geography to offer students an exciting introduction to the subject. At the college level, many geography departments are small; they are absent altogether at some high-profile universities.

Perhaps the greatest challenge is to overcome the public's view of geography as a simple exercise in place-name recognition. Much of geography's power lies in the insights it sheds on the nature and meaning of the evolving spatial arrangements and landscapes that make up our world. The importance of those insights should not be underestimated at a time of changing political boundaries, accelerated human alteration of the environment, and rapidly shifting patterns of human interaction.

Alexander B. Murphy is a professor and head of the geography department at the University of Oregon, and a vice-president of the American Geographical Society.

Originally appeared in *The Chronicle of Higher Education*, October 30, 1998, p. 54. © 1998 by Alexander B. Murphy. Reprinted by permission of the author.

The Four Traditions of Geography

William D. Pattison

Late Summer, 1990

To Readers of the *Journal of Geography:*

I am honored to be introducing, for a return to the pages of the *Journal* after more than 25 years, "The Four Traditions of Geography," an article which circulated widely, in this country and others, long after its initial appearance—in reprint, in xerographic copy, and in translation. A second round of life at a level of general interest even approaching that of the first may be too much to expect, but I want you to know in any event that I presented the paper in the beginning as my gift to the geographic community, not as a personal property, and that I re-offer it now in the same spirit.

In my judgment, the article continues to deserve serious attention—perhaps especially so, let me add, among persons aware of the specific problem it was intended to resolve. The background for the paper was my experience as first director of the High School Geography Project (1961–63)—not all of that experience but only the part that found me listening, during numerous conference sessions and associated interviews, to academic geographers as they responded to the project's invitation to locate "basic ideas" representative of them all. I came away with the conclusion that I had been witnessing not a search for consensus but rather a blind struggle for supremacy among honest persons of contrary intellectual commitment. In their dialogue, two or more different terms had been used, often unknowingly, with a single reference, and no less disturbingly, a single term had been used, again often unknowingly, with two or more different references. The article was my attempt to stabilize the discourse. I was proposing a basic nomenclature (with explicitly associated ideas) that would, I

trusted, permit the development of mutual comprehension **and** confront all parties concerned with the pluralism inherent in geographic thought.

This intention alone could not have justified my turning to the NCGE as a forum, of course. The fact is that from the onset of my discomfiting realization I had looked forward to larger consequences of a kind consistent with NCGE goals. As finally formulated, my wish was that the article would serve "to greatly expedite the task of maintaining an alliance between professional geography and pedagogical geography and at the same time to promote communication with laymen" (see my fourth paragraph). I must tell you that I have doubts, in 1990, about the acceptability of my word choice, in saying "professional," "pedagogical," and "layman" in this context, but the message otherwise is as expressive of my hope now as it was then.

I can report to you that twice since its appearance in the *Journal*, my interpretation has received more or less official acceptance—both times, as it happens, at the expense of the earth science tradition. The first occasion was Edward Taaffe's delivery of his presidential address at the 1973 meeting of the Association of American Geographers (see *Annals AAG*, March 1974, pp. 1–16). Taaffe's working-through of aspects of an interrelation among the spatial, area studies, and man-land traditions is by far the most thoughtful and thorough of any of which I am aware. Rather than fault him for omission of the fourth tradition, I compliment him on the grace with which he set it aside in conformity to a meta-epistemology of the American university which decrees the integrity of the social sciences as a consortium in their own right. He was sacrificing such holistic

claims as geography might be able to muster for a freedom to argue the case for geography as a social science.

The second occasion was the publication in 1984 of *Guidelines for Geographic Education: Elementary and Secondary Schools*, authored by a committee jointly representing the AAG and the NCGE. Thanks to a recently published letter (see *Journal of Geography*, March-April 1990, pp. 85–86), we know that, of five themes commended to teachers in this source,

> The committee lifted the human environmental interaction theme directly from Pattison. The themes of place and location are based on Pattison's spatial or geometric geography, and the theme of region comes from Pattison's area studies or regional geography.

Having thus drawn on my spatial, area studies, and man-land traditions for four of the five themes, the committee could have found the remaining theme, movement, there too—in the spatial tradition (see my sixth paragraph). However that may be, they did not avail themselves of the earth science tradition, their reasons being readily surmised. Peculiar to the elementary and secondary schools is a curriculum category framed as much by theory of citizenship as by theory of knowledge: the social studies. With admiration, I see already in the committee members' adoption of the theme idea a strategy for assimilation of their program to the established repertoire of social studies practice. I see in their exclusion of the earth science tradition an intelligent respect for social studies' purpose.

Here's to the future of education in geography: may it prosper as never before.

W. D. P., 1990

Reprinted from the Journal of Geography, 1964, pp. 211–216.

In 1905, one year after professional geography in this country achieved full social identity through the founding of the Association of American Geographers, William Morris Davis responded to a familiar suspicion that geography is simply an undisciplined "omnium-gatherum" by describing an approach that as he saw it imparts a "geographical quality" to some knowledge and accounts for the absence of the quality elsewhere.[1] Davis spoke as president of the AAG. He set an example that was followed by more than one president of that organization. An enduring official concern led the AAG to publish, in 1939 and in 1959, monographs exclusively devoted to a critical review of definitions and their implications.[2]

Every one of the well-known definitions of geography advanced since the founding of the AAG has had its measure of success. Tending to displace one another by turns, each definition has said something true of geography.[3] But from the vantage point of 1964, one can see that each one has also failed. All of them adopted in one way or another a monistic view, a singleness of preference, certain to omit if not to alienate numerous professionals who were in good conscience continuing to participate creatively in the broad geographic enterprise.

The thesis of the present paper is that the work of American geographers, although not conforming to the restrictions implied by any one of these definitions, has exhibited a broad consistency, and that this essential unity has been attributable to a small number of distinct but affiliated traditions, operant as binders in the minds of members of the profession. These traditions are all of great age and have passed into American geography as parts of a general legacy of Western thought. They are shared today by geographers of other nations.

There are four traditions whose identification provides an alternative to the competing monistic definitions that have been the geographer's lot. The resulting pluralistic basis for judgment promises, by full accommodation of what geographers do and by plain-spoken representation thereof, to greatly expedite the task of maintaining an alliance between professional geography and pedagogical geography and at the same time to promote communication with laymen. The following discussion treats the traditions in this order: (1) a spatial tradition, (2) an area studies tradition, (3) a man-land tradition and (4) an earth science tradition.

Spatial Tradition

Entrenched in Western thought is a belief in the importance of spatial analysis, of the act of separating from the happenings of experience such aspects as distance, form, direction and position. It was not until the 17th century that philosophers concentrated attention on these aspects by asking whether or not they were properties of things-in-themselves. Later, when the 18th century writings of Immanuel Kant had become generally circulated, the notion of space as a category including all of these aspects came into widespread use. However, it is evident that particular spatial questions were the subject of highly organized answering attempts long before the time of any of these cogitations. To confirm this point, one need only be reminded of the compilation of elaborate records concerning the location of things in ancient Greece. These were records of sailing distances, of coastlines and of landmarks that grew until they formed the raw material for the great *Geographia* of Claudius Ptolemy in the 2nd century A.D.

A review of American professional geography from the time of its formal organization shows that the spatial tradition of thought had made a deep penetration from the very beginning. For Davis, for Henry Gannett and for most if not all of the 44 other men of the original AAG, the determination and display of spatial aspects of reality through mapping were of undoubted importance, whether contemporary definitions of geography happened to acknowledge this fact or not. One can go further and, by probing beneath the art of mapping, recognize in the behavior of geographers of that time an active interest in the true essentials of the spatial tradition— *geometry* and *movement*. One can trace a basic favoring of movement as a subject of study from the turn-of-the-century work of Emory R. Johnson, writing as professor of transportation at the University of Pennsylvania, through the highly influential theoretical and substantive work of Edward L. Ullman during the past 20 years and thence to an article by a younger geographer on railroad freight traffic in the U.S. and Canada in the *Annals* of the AAG for September 1963.[4]

One can trace a deep attachment to geometry, or positioning-and-layout, from articles on boundaries and population densities in early 20th century volumes of the *Bulletin of the American Geographical Society*, through a controversial pronouncement by Joseph Schaefer in 1953 that granted geographical legitimacy only to studies of spatial patterns[5] and so onward to a recent *Annals* report on electronic scanning of cropland patterns in Pennsylvania.[6]

One might inquire, is discussion of the spatial tradition, after the manner of the remarks just made, likely to bring people within geography closer to an understanding of one another and people outside geography closer to an understanding of geographers? There seem to be at least two reasons for being hopeful. First, an appreciation of this tradition allows one to see a bond of fellowship uniting the elementary school teacher, who attempts the most rudimentary instruction in directions and mapping, with the contemporary research geographer, who dedicates himself to an exploration of central-place theory. One cannot only open the eyes of many teachers to the potentialities of their own instruction, through proper exposition of the spatial tradition, but one can also "hang a bell" on research quantifiers in geography, who are often thought to have wandered so far in their intellectual adventures as to have become lost from the rest. Looking outside geography, one may anticipate benefits from the readiness of countless persons to associate the name "geography" with maps. Latent within this readiness is a willingness to recognize as geography, too, what maps are about—and that is the geometry of and the movement of what is mapped.

Area Studies Tradition

The area studies tradition, like the spatial tradition, is quite strikingly represented in classical antiquity by a practitioner to whose surviving work we can point. He is Strabo, celebrated for his *Geography* which is a massive production addressed to the statesmen of Augustan Rome and intended to sum up and regularize knowledge not of the location of places and associated cartographic facts, as in the somewhat later case of Ptolemy, but of the nature of places, their character and their differentiation. Strabo exhibits interesting attributes of the area-studies tradition that can hardly be overemphasized. They are a pronounced tendency toward subscription primarily to literary standards, an almost omnivorous appetite for information and a self-conscious companionship with history.

It is an extreme good fortune to have in the ranks of modern American geography the scholar Richard Hartshorne, who has pondered the meaning of the area-studies tradition with a legal acuteness that few persons would challenge. In his *Nature of Geography*, his 1939 monograph already cited,[7] he scrutinizes exhaustively the implications of the "interesting attributes" identified in connection with Strabo, even though his concern is with quite other and much later authors, largely German. The major literary problem of unities or wholes he considers from every angle. The Gargantuan appetite for miscellaneous information he accepts and rationalizes. The companionship between area studies and history he clarifies by appraising the so-called idiographic content of both and by affirming the tie of both to what he and Sauer have called "naively given reality."

The area-studies tradition (otherwise known as the chorographic tradition) tended to be excluded from early American professional geography. Today it is beset by certain champions of the spatial tradition who would have one believe that somehow the area-studies way of organizing knowledge is only a subdepartment of spatialism. Still, area-studies as a method of presentation lives and prospers in its own right. One can turn today for reassurance on this score to practically any issue of the *Geographical Review*, just as earlier readers could turn at the opening of the century to that magazine's forerunner.

What is gained by singling out this tradition? It helps toward restoring the faith of many teachers who, being accustomed to administering learning in the area-studies style, have begun to wonder if by doing so they really were keeping in touch with professional geography. (Their doubts are owed all too much to the obscuring effect of technical words attributable to the very professionals who have been intent, ironically, upon protecting that tradition.) Among persons outside the classroom the geographer stands to gain greatly in intelligibility. The title "area-studies" itself carries an understood message in the United States today wherever there is contact with the usages of the academic community. The purpose of characterizing a place, be it neighborhood or nation-state, is readily grasped. Furthermore, recognition of the

right of a geographer to be unspecialized may be expected to be forthcoming from people generally, if application for such recognition is made on the merits of this tradition, explicitly.

Man-Land Tradition

That geographers are much given to exploring man-land questions is especially evident to anyone who examines geographic output, not only in this country but also abroad. O. H. K. Spate, taking an international view, has felt justified by his observations in nominating as the most significant ancient precursor of today's geography neither Ptolemy nor Strabo nor writers typified in their outlook by the geographies of either of these two men, but rather Hippocrates, Greek physician of the 5th century B.C. who left to posterity an extended essay, *On Airs, Waters and Places*.[8] In this work made up of reflections on human health and conditions of external nature, the questions asked are such as to confine thought almost altogether to presumed influence passing from the latter to the former, questions largely about the effects of winds, drinking water and seasonal changes upon man. Understandable though this uni-directional concern may have been for Hippocrates as medical commentator, and defensible as may be the attraction that this same approach held for students of the condition of man for many, many centuries thereafter, one can only regret that this narrowed version of the man-land tradition, combining all too easily with social Darwinism of the late 19th century, practically overpowered American professional geography in the first generation of its history.[9] The premises of this version governed scores of studies by American geographers in interpreting the rise and fall of nations, the strategy of battles and the construction of public improvements. Eventually this special bias, known as environmentalism, came to be confused with the whole of the man-land tradition in the minds of many people. One can see now, looking back to the years after the ascendancy of environmentalism, that although the spatial tradition was asserting itself with varying degrees of forwardness, and that although the area-studies tradition was also making itself felt, perhaps the most interesting chapters in the story of American professional geography were being written by academicians who were reacting against environmentalism while deliberately remaining within the broad man-land tradi-

tion. The rise of culture historians during the last 30 years has meant the dropping of a curtain of culture between land and man, through which it is asserted all influence must pass. Furthermore work of both culture historians and other geographers has exhibited a reversal of the direction of the effects in Hippocrates, man appearing as an independent agent, and the land as a sufferer from action. This trend as presented in published research has reached a high point in the collection of papers titled *Man's Role in Changing the Face of the Earth*. Finally, books and articles can be called to mind that have addressed themselves to the most difficult task of all, a balanced tracing out of interaction between man and environment. Some chapters in the book mentioned above undertake just this. In fact the separateness of this approach is discerned only with difficulty in many places; however, its significance as a general research design that rises above environmentalism, while refusing to abandon the man-land tradition, cannot be mistaken.

The NCGE seems to have associated itself with the man-land tradition, from the time of founding to the present day, more than with any other tradition, although all four of the traditions are amply represented in its official magazine, *The Journal of Geography* and in the proceedings of its annual meetings. This apparent preference on the part of the NCGE members *for defining geography in terms of the man-land tradition* is strong evidence of the appeal that man-land ideas, separately stated, have for persons whose main job is teaching. It should be noted, too, that this inclination reflects a proven acceptance by the general public of learning that centers on resource use and conservation.

Earth Science Tradition

The earth science tradition, embracing study of the earth, the waters of the earth, the atmosphere surrounding the earth and the association between earth and sun, confronts one with a paradox. On the one hand one is assured by professional geographers that their participation in this tradition has declined precipitously in the course of the past few decades, while on the other one knows that college departments of geography across the nation rely substantially, for justification of their role in general education, upon curricular content springing directly from this tradition. From all the reasons that combine to account for this state of affairs, one may, by selecting only two, go far toward achieving an understanding of this tradition. First, there is the fact that American college geography, growing out of departments of geology in many crucial instances, was at one time greatly overweighted in favor of earth science, thus rendering the field unusually liable to a sense of loss as better balance came into being. (This one-time disproportion found reciprocate support for many years in the narrowed, environmentalistic interpretation of the man-land tradition.) Second, here alone in earth science does one encounter subject matter in the normal sense of the term as one reviews geographic traditions. The spatial tradition abstracts certain aspects of reality; area studies is distinguished by a point of view; the man-land tradition dwells upon relationships; but earth science is identifiable through concrete objects. Historians, sociologists and other academicians tend not only to accept but also to ask for help from this part of geography. They readily appreciate earth science as something physically associated with their subjects of study, yet generally beyond their competence to treat. From this appreciation comes strength for geography-as-earth-science in the curriculum.

Only by granting full stature to the earth science tradition can one make sense out of the oft-repeated addage, "Geography is the mother of sciences." This is the tradition that emerged in ancient Greece, most clearly in the work of Aristotle, as a wide-ranging study of natural processes in and near the surface of the earth. This is the tradition that was rejuvenated by Varenius in the 17th century as "Geographia Generalis." This is the tradition that has been subjected to subdivision as the development of science has approached the present day, yielding mineralogy, paleontology, glaciology, meterology and other specialized fields of learning.

Readers who are acquainted with American junior high schools may want to make a challenge at this point, being aware that a current revival of earth sciences is being sponsored in those schools by the field of geology. Belatedly, geography has

joined in support of this revival.[10] It may be said that in this connection and in others, American professional geography may have faltered in its adherence to the earth science tradition but not given it up.

In describing geography, there would appear to be some advantages attached to isolating this final tradition. Separation improves the geographer's chances of successfully explaining to educators why geography has extreme difficulty in accommodating itself to social studies programs. Again, separate attention allows one to make understanding contact with members of the American public for whom surrounding nature is known as the geographic environment. And finally, specific reference to the geographer's earth science tradition brings into the open the basis of what is, almost without a doubt, morally the most significant concept in the entire geographic heritage, that of the earth as a unity, the single common habitat of man.

An Overview

The four traditions though distinct in logic are joined in action. One can say of geography that it pursues concurrently all four of them. Taking the traditions in varying combinations, the geographer can explain the conventional divisions of the field. Human or cultural geography turns out to consist of the first three traditions applied to human societies; physical geography, it becomes evident, is the fourth tradition prosecuted under constraints from the first and second traditions. Going further, one can uncover the meanings of "systematic geography," "regional geography," "urban geography," "industrial geography," etc.

It is to be hoped that through a widened willingness to conceive of and discuss the field in terms of these traditions, geography will be better able to secure the inner unity and outer intelligibility to which reference was made at the opening of this paper, and that thereby the effectiveness of geography's contribution to American education and to the general American welfare will be appreciably increased.

Notes

1. William Morris Davis, "An Inductive Study of the Content of Geography," *Bulletin of the American Geographical Society*, Vol. 38, No. 1 (1906), 71.
2. Richard Hartshorne, *The Nature of Geography*, Association of American Geographers (1939), and idem., *Perspective on the Nature of Geography*, Association of American Geographers (1959).
3. The essentials of several of these definitions appear in Barry N. Floyd, "Putting Geography in Its Place," *The Journal of Geography*, Vol. 62, No. 3 (March, 1963), 117–120.
4. William H. Wallace, "Freight Traffic Functions of Anglo-American Railroads," *Annals of the Association of American Geographers*, Vol. 53, No. 3 (September, 1963), 312–331.
5. Fred K. Schaefer, "Exceptionalism in Geography: A Methodological Examination," *Annals of the Association of American Geographers*, Vol. 43, No. 3 (September, 1953), 226–249.
6. James P. Latham, "Methodology for an Instrumental Geographic Analysis," *Annals of the Association of American Geographers*, Vol. 53, No. 2 (June, 1963), 194–209.
7. Hartshorne's 1959 monograph, *Perspective on the Nature of Geography*, was also cited earlier. In this later work, he responds to dissents from geographers whose preferred primary commitment lies outside the area studies tradition.
8. O. H. K. Spate, "Quantity and Quality in Geography," *Annals of the Association of American Geographers*, Vol. 50, No. 4 (December, 1960), 379.
9. Evidence of this dominance may be found in Davis's 1905 declaration: "Any statement is of geographical quality if it contains… some relation between an element of inorganic control and one of organic response" (Davis, *loc. cit.*).
10. Geography is represented on both the Steering Committee and Advisory Board of the Earth Science Curriculum Project, potentially the most influential organization acting on behalf of earth science in the schools.

From *Journal of Geography*, September/October 1990, pp. 202–206. © 1990 by the National Council for Geographic Education. Reprinted by permission.

The Power of Place

The concept of place-based education shapes Sharon Bishop's English language arts curriculum. High school students read literature of the region, inquire about the area they live in, and write the stories of their community. In the process, they come to recognize the value of community and acquire the skills to "live well anywhere."

Sharon Bishop

Where are you from? is a frequently asked question in the rural Midwest where I live. A place identifies us. As a longtime teacher of high school English in mid-central Nebraska, I am convinced that understanding how we connect to our place is a crucial element in our ability to live well as individuals and as communities. In my secondary class rooms, I have tried to teach this understanding of place through a study of regional literature, through community inquiry, and through a connection to nature. The curriculum has evolved over a number of years. Its growth has been supported by my work with the Nebraska Writing Project and School at the Center, an affiliate of the Annenberg Rural Challenge.

For almost a decade, my colleagues and I in the Nebraska Writing Project have worked to distill our approach to this kind of teaching and learning—place-based or place-conscious education. The Nebraska Writing Project is a leader in the development of a summer rural writing institute, for which I regularly serve as a facilitator. I was a member of a national project—Rural Voices, Country Schools—that documented, in six writing projects across the United States, exemplary education in rural schools. The Nebraska team has described our experiences in *Ritual Voices: Place-Conscious Education and the Teaching of Writing* (Brooke). I also served as coeditor for a publication of student writing from Nebraska rural students, All Roads Lead Home (Bishop). My experiences in the classroom and in the rural writing institutes reinforce the power of place. Place-based education helps us understand our lives and our work better. It is especially crucial to rural schools, those places that are disappearing because of population declines and school consolidations.

The Power of Community

In the small towns of rural Nebraska, a place is often defined by its public school. These small schools are the center of community life. The majority of citizens attend the students' games and performances, and the school often serves as a community center. In most cases, the name of the school and the name of the town are synonymous. However, in the last decade, de-clining populations have resulted in the consolidation of many small schools. When a community loses a school, it may also lose its identity. New school names surface: High Plains, Cross County, TriCounty. Some schools, such as the one where I teach, have created longer titles that honor the names of both towns. We are Heartland Community Schools—Henderson/Bradshaw. A real possibility exists that other community names may be incorporated into the title in future mergers.

When a community loses a school, it may also lose its identity.

In 1997, our school became part of a partnership to improve public education, the Annenberg Rural Challenge. The goal of the Challenge was

> to transform rural schools and communities.... Among the dominant themes are that students should come to know their local communities well, that communities see the schools and students as crucial assets and that communities and schools need to become more integrated. (*Living* 5)

In Nebraska, we formed a consortium called School at the Center. Eleven rural communities worked with staff from the University of Nebraska, Lincoln. Dr. Robert Brooke, director of the Nebraska Writing Project, described this work:

> The explicit purpose for School at the Center was to aid in the revitalization of rural communities through reimagining local schools as a centering force for place-conscious living. The program had five strands: region-centered humanities, sustainable agriculture and regional biological awareness, entrepreneurship training, development of sustainable local housing, and region-centered math/science education. These strands required cooperation between schools and civic leaders in each community. (15)

The Pedagogy of Place

I became associated with School at the Center in part because I had constructed a curriculum twenty years ago that embraced many of these ideas. For two decades, I have used the idea of place as the centerpiece of a sophomore English class, although the term *place-conscious education* was then unknown to me. Twenty years ago I received permission from my administrators to discard a traditional anthology and substitute the work of Nebraska authors. I chose Willa Cather, John G. Neihardt, Mari Sandoz, and Bess Streeter Aldrich. Their works reflect the early settlement history of Nebraska, and the authors often describe the prairie that the first settlers found—huge spaces open to an endless sky. The physical landscape and the stories of the immigrants who made new homes here shaped their writing. This literature remains in the curriculum.

I chose *stories* as the centerpiece that would direct our study of place; after all, the fiction of Cather and other Nebraska writers is bound up in the local.

Over the years, the curriculum incorporated more material from the local setting. The 1997 affiliation with the Annenberg Rural Challenge and School at the Center introduced me to a formal definition of *pedagogy of place*:

> In its most simple form, pedagogy/curriculum of place is an expression of the growing recognition of context and locale and their unique contributions to the educational project. Using what is local and immediate as a source of curriculum tends to deepen knowledge through the larger understandings of the familiar and accessible. It clearly increases student understanding and often gives a stronger impetus to apply problem-solving skills....
>
> A pedagogy of place, then, recontextualizes education locally. It makes education a preparation for citizenship, both locally and in wider contexts, while also providing the basis for continuing scholarship. (*Living* 11–12)

My association with School at the Center confirmed what I had been doing in my classroom. I also learned that place-based work does nor have a fixed definition. That is one of its beauties: It is specific to the locale, or place, that produces it.

Gregory Smith identified five thematic patterns in place-based education. All can be adapted to specific settings:

- Cultural Studies
- Nature Studies
- Real-World Problem Solving
- Internships and Entrepreneurial
- Opportunities
- Induction into Community Processes

My curriculum combines two of these patterns: Cultural Studies and Nature Studies. I chose *stories* as the centerpiece that would direct our study of place; after all, the fiction of Cather and other Nebraska writers is bound up in the local. Reading would lead us to exploring our area through research, interviews, writing, and photography. The sources would be the students, community, and land.

In *What a Writer Needs*, Ralph Fletcher stresses the importance of place as a starting point for writers:

> I usually start writing with something I know: a detail, an image, a snatch of overheard conversation, a story, a person. A place. Place is an excellent starring point because places live in the deepest parts of us. In one sense, we never leave them: We soak them up, carry them around, all the various places we have known. (114)

Collecting the Stories of This Place

The community where I teach is a stable one, with many generations of families living in the area. Family celebrations are common. Students are aware of the early settlement history: Churches and local historical groups have written documentations of the first immigrant house built in 1874 for the Germans from Russia who came to Nebraska on the Burlington Railroad. I wanted students to collect more recent stories: the terrible dust storms of the "Dirty Thirties" and the innovative deep-well drilling that started in this area and irrigated a dry land; the building of Interstate 80 in the 1950s; and memories of the cold war. I also wanted to include the family stories that make up the fabric of this place. More importantly, I wanted students to discover that the people who live in this small community, which often seems so boring to them, have had vibrant lives of accomplishment and sadness, lives of large and small victories as well as failures.

Students brought rich stories to the classroom. One commented, "I heard stories just like the ones in *My Ántonia!*" Meaningful poetry and essays based on the collected stories go into a class-made booklet that will be read by family and community members. Students come to see the members of their families and community as real people whose lives have made significant contributions to this place.

Paul Gtuchow writes about the power and importance of family and community stories:

> All history is ultimately local and personal. To tell what we remember, and to keep on telling it, is to keep the past alive in the present. Should we not do so, we could not know, in the deepest sense, how to inhabit a place. To inhabit a place means literally to have made it a habit, to have made it the custom and ordinary practice of our lives.... We own places not because we possess the deeds to them, but because they have entered the continuum of our lives. (6)

When students write "Where I'm From" poems, they see what Gruchow means. The idea comes from a poem that George Ella Lyon included in *Where I'm From: Where Poems Come From*. "Where I'm From" shows students that the ordinary foods, customs, and keepsakes of a family can create personal poetry of place (3). Some students follow Lyon's structure; others use the basic idea to create their own poetry. This pattern is a powerful tool and the students and their parents value the writing that comes from it.

I am from Kool-Aid stands and sugar cookies.
From dodgeball and basketball.... (Emily)
At the end of my parent's bed a cedar chest
full of old dresses
from years gone by,
some worn only a few times.
I'm from those dresses—
dolls to weddings—
sewn for special occasions
that stitch the family together. (Danelle)

Writings based on oral heritage interviews connect students to their families and to the elders of the community:

The bright sun finally rose on the morning of March 12. Darlene had been waiting all month for this day to come! It was her eighth birthday, and she was more than excited. She jumped out of her feather bed, crossing her fingers in hopes for a new horse. (Dani)

The teacher turned to write on the board, and that was our chance for the mischievous act we were about to perform.
"One, two, three!" we whispered.
The one-room schoolhouse was just like any other in the 1930s. (Lindsey)

When I asked my grandma about a special childhood memory or toy, she smiled brightly and said one word-"Anna!" This is the story of Anna and my grandma. (Sara)

I lay in bed, inches away from my sleeping twin sisters. I felt as though I could hardly stand still. It was the 4th of July 1939. My 6 sisters and 3 brothers eagerly waited for this day each year. (Emily) His father drove him to school. It was a new environment, a place to make new friends. And it was time to learn English. LeRoy's family always spoke German at home. He had only heard English a few times.... (Ashley)

With a long week of work and a full tank of gas Ardean and his '55 Chevy were ready for a night on the town. (Todd)

Students come to see the members of their families and community as real people whose lives have made significant contributions to this place.

Stewardship of Place

An integral part of this curriculum of place is to give students opportunities to become knowledgeable about their natural environment. We often assume that rural students have this knowledge, but that is not always true. Agricultural and technological advances have removed people from the close association with the land that previous generations experienced. My students who live on farms do not do the chores of their parents and grandparents; they do not milk cows or gather eggs.

The formal literature also leads us outside of the classroom. Because the prairie forms the setting for the novels that we read, a study of this ecosystem helps students understand some of the

difficulties the first settlers faced. The science teacher and I coordinate a visit to a preserved prairie of forty acres that represents how all of the state once looked. Field trips to a protected wetland and to the Platte River when the sandhill cranes visit each spring lead students to consider stewardship of place. In these outdoor classrooms, we pay attention to the land. Students discover for themselves the ways that humans have changed this ecosystem.

Field trips to a protected wetland and to the Platte River when the sandhill cranes visit each spring lead students to consider stewardship of place.

They also observe that those who have formed a community in a place will care for it. In my classroom, students discuss issues that are both local and global, such as water use. No lectures can replace the experience of seeing the old banks of the Platte River several miles away from a present-day river that has been diverted and changed by the humans who live here. Some students may be skeptical about the protection afforded the sandhill cranes that land here on the Platte for six weeks every spring. But on a chilly March evening, as we huddle in a blind at sunset and listen to the cries of thousands of cranes filling up the riverbanks, a new understanding is awakened in the students. These outdoor classrooms not only offer rich opportunities for writing, but they also allow students to discover that living in a place requires stewardship of that place.

When students know that a piece of writing represents a story entrusted to them in an oral heritage interview, they are more careful writers.

After our visits to these outdoor classrooms, students write:

As we soar home / I am aware that the sophomore and crane migrations are similar / We are both soaring through life, / Looking to land but moving too fast. (Tim)

The Platte is dark / we sit blinded in the blind / deafened as we listen. (Gram)

I lie down by the foxtail and the musk thistle / comforted by their presence / knowing they still exist. (Nikki)

The Importance of Teaching Place

Work that is closely related to a local place and the people who live there allows students to show what they have learned in a number of ways. The literature that we read, the stories collected from family and community, and the stories from the places we visit allow students to develop critical-thinking skills. They use these resources to create poetry, prose, and photography. They teach one another through group experiments at the prairie and wetland and through presentations based on that work. They understand that they will have an audience beyond the classroom for the created products. We regularly create class-published books and present displays and programs for the community. When students know that a piece of writing represents a story entrusted to them in an oral heritage interview, they are more careful writers. This is challenging, but authentic, work.

Place-conscious education also allows students to learn to value a small town that seems boring. They are often astounded by the wisdom and experiences of elders in the community. Perhaps they will participate in solving the problem of dwindling populations in the heartland. A stretch of prairie or a view of the Platte River at sunset offers them an aesthetic value for this location. Students describe these field trips as peaceful, stress-free places. They may become future preservationists of these environments.

There is power in the teaching of place. School at the Center asked me to help students value the rural places where they live so that they would return to them, thereby halting the loss of rural towns and schools in Nebraska. I do not know if this is an outcome of the place-conscious curriculum that I teach. Some former students have returned to live in this community. Is it because of the work that was done in my class? I do not know; perhaps that played some small part in the decision. Still, if students are allowed to learn how to care about a place and to care for it, they are more likely to consider living there and helping to solve its problems. A pride of place will also give them the necessary skills to live well in any community. Place-based learning, wherever that place is, teaches a sense of community and gives students a model for living well anywhere.

References

Bishop, Sharon, Carol McDaniels, and Miles Bryant, eds. *All Roads Lead Home*. Lincoln: Dageforde, 2001.

Brooke, Robert E. "Place-Conscious Education, Rural Schools, and the Nebraska Writing Project's Rural Voices, Country Schools Team." *Rural Voices: Place-Conscious Education and the Teaching of Writing*. Ed. Robert E. Brooke. New York: Teachers College, 2003. 1–19.

Fletcher, Ralph. *What a Writer Needs*. Pottsmouth: Heinemann, 1993.

Gruchow, Paul. *Grass Roots: The Universe of Home*. Minneapolis: Milkweed, 1995.

Living and Learning in Rural Schools and Communities: A Report to the Annenberg Rural Challenge. Cambridge: Harvard Graduate School of Education, 1999.

Lyon, George Ella. *Where I'm From: Where Poems Come From*. Spring: Absey, 1999.

Smith, Gregory A. "Place-Based Education: Learning to Be Where We Are." *Phi Delta Kappan* 83.8 (2002): 584–94

A veteran secondary English teacher in Henderson, Nebraska, **Sharon Bishop** is co-director of the Nebraska Writing Project and has facilitated several summer Rural Writing Institutes. *email:* sbishop@esu6.org.

The Changing Landscape of Fear

SUSAN L. CUTTER, DOUGLAS B. RICHARDSON,
AND THOMAS J. WILBANKS

In the days following September 11, 2001, all geographers felt a sense of loss—people we knew perished, and along with everyone else we experienced discomfort in our own lives and a diminished level of confidence that the world will be a safe and secure place for our children and grandchildren. Many of us who are geographers felt an urge and a need to see if we could find ways to apply our knowledge and expertise to make the world more secure. A number of our colleagues assisted immediately by sharing specific geographical knowledge (such as Jack Shroder's expert knowledge on the caves in Afghanistan) or more generally by assisting rescue and relief efforts through our technical expertise in Geographic Information System (GIS) and remote sensing (such as Hunter College's Center for the Analysis and Research of Spatial Information and various geographers at federal agencies and in the private sector). Still others sought to enhance the nation's research capacity in the geographical dimensions of terrorism (the Association of American Geographers' Geographical Dimensions of Terrorism project). Many of us have given considerable thought to how our science and practice might be useful in both the short and longer terms. One result is the set of contributions to this book.

But, we fail in our social responsibility if we spend our time thinking of geography as the end. Geography is not the end; it is one of many means to the end. Our concern should be with issues and needs that transcend any one discipline. As we address issues of terrorism, utility without quality is unprofessional, but quality without utility is self-indulgent. Our challenge is to focus not on geography's general importance but on the central issues in addressing terrorism as a new reality in our lives in the United States (although, unfortunately, not a new issue in too many other parts of our world).

The September 11, 2001 events have prompted both immediate and longer-term concerns about the geographical dimensions of terrorism. Potential questions on the very nature of these types of threats, how the public perceives them, individual and societal willingness to reduce vulnerability to such threats, and ultimately our ability to manage their consequences require concerted research on the part of the geographical community, among others. Geographers are well positioned to address some of the initial questions regarding emergency management and response and some of the spatial impacts of the immediate consequences, but the research community is not sufficiently mobilized and networked internally or externally to develop a longer, sustained, and theoretically informed research agenda on the geographical dimensions of terrorism. As noted more than a decade ago, "issues of nuclear war and deterrence [and now terrorism] are inherently geographical, yet our disciplinary literature is either silent on the subject or poorly focused" (Cutter 1988: 132). Recent events provide an opportunity and a context for charting a new path to bring geographical knowledge and skills to the forefront in solving this pressing international problem.

PROMOTING LANDSCAPES OF FEAR

Terrorists (and terrorism) seek to exploit the everyday—things that people do, places that they visit, the routines of daily living, and the functioning of institutions. Terrorism is an adaptive threat which changes its target, timing, and mode of delivery as circumstances are altered. The seeming randomness of terrorist attacks (either the work of organized groups or renegade individuals) in both time and space increases public anxiety concerning terrorism. At the most fundamental level, September 11, 2001 was an attack on the two most prominent symbols of U.S. financial and military power: the World Trade Center and the Pentagon (Smith 2001, Harvey 2002). The events represented symbolic victories of chaos over order and normalcy (Alexander 2002), disruptions in and the undermining of global financial markets (Harvey 2002), a nationalization of terror (Smith 2002), and the creation of fear and uncertainty among the public, precisely the desired outcome by the perpetrators. In generating this psychological landscape of fear, people's activity patterns were and are being altered, with widespread social, political, and economic effects. The reduction in air travel by consumers in the weeks and months following September 11, 2001 was but one among many examples.

WHAT ARE THE FUNDAMENTAL ISSUES OF TERRORISM?

There are a myriad of different ways to identify and examine terrorism issues. Some of these dimensions are quite conventional, others less so. In all cases, geographical understanding provides an essential aspect of the inquiry. There are a number of dimensions of the issues that seem reasonably clear. For instance, one conventional way of looking at the topic is to distinguish four central subject-matter challenges:

1. *Reducing threats*, including a) reducing the reasons why people want to commit terrorist acts, thereby addressing root causes, and b) reducing the ability of potential terrorists to accomplish their aims, or deterrence.
2. *Detecting threats* that have not been avoided, using sensors and signature detection to spot potential actions before they happen and interrupt them.
3. *Reducing vulnerabilities to threats*, focusing on critical sectors and infrastructures, hopefully without sacrificing civil liberties and individual freedoms.
4. *Improving responses to terrorism*, emphasizing "consequence management," and also attributing causation and learning from experience (for example, forensics applied to explosive materials and anthrax strains).

A different way of viewing terrorism is according to time horizons. Immediately after September 11, 2001, governmental leaders told us that the nation was now engaged in a new "war on terrorism" that will last several years, and that our existing knowledge and technologies are needed for this war. Early estimates of the overall U.S. national effort are very large—in the range of $30 to $40 billion per year—including the formation of a new executive department, the Department of Homeland Security. Early priorities include securing national borders, supporting first responders mainly in the Federal Emergency Management Agency (FEMA) and the Department of Justice, defending against bioterrorism, and applying information technologies to improve national security.

Beyond this, we know that better knowledge and practices should be put to use in the next half decade or so, as we face a challenge that is more like a stubborn virus than a single serial killer. To address this type of need, attention often is placed on capabilities where progress can be made relatively quickly if resources are targeted carefully. Some of our CIS and GIScience tools are especially promising candidates for such enhancements, which have both positive and negative consequences (Monmo-

nier 2002). The use of such technologies surely will help secure homelands, but at what price, the loss of personal freedoms or invasion of privacy?

There are other dimensions as well. For instance, one dimension concerns boundaries between free exchanges of information and limited ones, between classified work and unclassified work. Another differentiates between different types of threats: physical violence, chemical or biological agents, cyberterrorism, and the like. Still other themes are woven through the material that follows.

THE CHALLENGE AHEAD

The greatest challenge to geographers and our colleagues in neighboring fields of study is to stretch our minds beyond familiar research questions and specializations so as to be innovative, even ingenious, in producing new understandings that contribute to increased global security. Clearly, the most serious specific threats to security in the future will be actions that are difficult to imagine now: social concerns just beginning to bubble to the surface, technologies yet to be developed, biological agents that do not yet exist, terrorist practices that are beyond our imagination. A core challenge is to improve knowledge and institutional capacities that prepare us to deal with the unknown and the unexpected, with constant change calling for staying one step ahead instead of always being one step behind. When research requires, say, three years to produce results and another two years to communicate in print to prospective audiences, we need to be unusually prescient as we construct our research agendas related to terrorism issues, and we need to be very perceptive and skillful in convincing non-geographers that these longer-term research objectives are, in fact, truly important.

The topic of combating terrorism is not an easy one. It calls for us to stretch in directions that may be new and not altogether comfortable. It threatens to entangle us in policy agendas that many of us may consider insensitively conceived, even distasteful. It may endanger social cohesion in our own community of scholars. On the other hand, how can we turn our backs on a phenomenon that threatens political freedom, social cohesion far beyond our own cohorts, economic progress, environmental sustainability, and many other values that we hold dear, including the future security of our own children and grandchildren?

More fundamentally, geographers are not concerned only with winning the war on terrorism in the next two years or deploying new capabilities in the next five or ten. We are concerned with working toward a secure century, restoring a widespread sense of security in the global society in the longer term without undermining basic freedoms. This is the domain of the research world; assuring a stream of new knowledge, understandings, and tools for the longer term, and looking for policies and practices that—if they could be conceived and used—would make a significant difference in the quality of life.

As we prepare to create this new knowledge and understandings, what we are trying to do, in fact, is to create the new twenty-first-century utility—not a hardened infrastructure such as for power or water, but rather a geographical understanding and spatial infrastructure that helps the nation understand and respond to threats. The effort required to create this new utility to serve the nation has an historical analogy in the creation of the Tennessee Valley Authority (TVA), under Franklin Roosevelt's New Deal. The Appalachian region the southeastern United States had a long history of economic depression and was among those areas hardest hit by the Great Depression of the 1930s. The creation of the TVA, a multipurpose utility with an economic development mission, constructed dams for flood control and hydroelectric power for the region in order to: 1) bring electricity to the rural areas that did not have it; 2) stimulate new industries to promote economic development; 3) control flooding, which routinely plagued the region; and 4) develop a more sustainable and equitable future for the region's residents. This twenty-first-century utility must rely on geographical knowledge and synthesis capabilities as we begin to understand the root causes of insecurity both here and abroad, vulnerabilities and resiliencies in our daily lives and the systems that support them, and our collective role in fostering a more sustainable future, both domestically and globally.

Much of the content of this book is aimed at this longer term, and it is important for geography to join with others in the research community to assure that the long term is not neglected as research support is directed toward combating terrorism and protecting homelands in the short run. This is why the Association of American Geographers and some of its members have joined together to produce the perspectives and insights represented in this book. It is only a start, we still have a long way to go, and there are daunting intellectual and political hazards to be overcome. But if many of us will keep a part of our professional focus on this global and national issue, we have a chance to make our world better in many tangible ways.

Watching over the world's oceans

A quick technological fix is not the best response to the December tsunami.

Keith Alverson

In the mid-nineteenth century, the HMS Beagle docked in Concepciòn, Chile, giving Charles Darwin the opportunity to see and describe the immediate aftermath of a tidal wave. His eyewitness account in the classic *Voyage of the Beagle* could easily be read as a report from Sri Lanka after the tsunami of 26 December 2004. The timeless nature of the devastation stands in stark contrast to the enormous progress that has occurred since then in relevant areas of science, technology and intergovernmental cooperation—progress that should have made a difference. Plate tectonics, accurate seafloor mapping, powerful computer calculations for wave propagation, real-time wireless global communications networks and operational 24-hour government warning systems are all new since Darwin's time. It seems they made no difference. With hindsight, they could have, and should have.

The December tsunami was a natural catastrophe, but much of the death and destruction that followed was a result of the collective failure of human institutions. Not surprisingly, hindsight has informed the global response. In addition to the outpouring of aid, there is interest from nations wishing to build an operational tsunami warning system in the Indian Ocean as soon as possible. Although laudable, this goal is far too narrow. Why? Despite local tsunamis being a frequent occurence in the Indian ba-

sin, we have no idea when or where to expect the next large regional tsunami. It could be centuries away. A rapidly developed, single-basin, single-purpose tsunami warning system that goes unused for many years is likely to be falling apart by the time it is called to use.

This is not a wholly pessimistic view—we have been here before. Following two major tsunamis in the Pacific in the early 1960s, the Intergovernmental Oceanographic Commission of UNESCO (IOC) and its member states set up a warning system for that ocean. By 2004, the funding for the upkeep of that system was a trickle, and three of its six seafloor pressure sensors were out of commission. There has long been talk of expanding and upgrading the Pacific warning system, which lacks regional tsunami warning centres in many vulnerable areas—southeast Asia, the southwest Pacific, and Central and South America. Unfortunately, once the initial system was in place, the resources required to maintain it properly—let alone expand or improve it—were extremely difficult to find. Building a single-use warning system for the Pacific basin alone in response to the events of the early 1960s was arguably not the best thing to do. It would be a mistake for the international scientific community to suggest another quick technological fix for the Indian ocean, where tsunamis are even less frequent.

A multihazard approach

A more sensible idea is to develop a global tsunami warning system that is fully integrated with an operational ocean-observing system—one that is regularly used for other related hazards, such as storm surges. Storm surges associated with tropical cyclones can hit coastal areas well ahead of the landfall of the actual storm; they travel with nearly the same rapidity as tsunamis, but occur much more frequently. And for unprepared or unwarned populations, they can be equally deadly. For example, in 1970 (and again in 1991) six- to seven-metre-high storm surges striking Bangladesh resulted in around half a million deaths. At present, there is no regional system for predicting storm surges, although there are a few national efforts. But tide gauges provide vital information for the high-resolution models used in storm-surge prediction—and these are the same data needed for tsunami warnings.

Although the scope of a tsunami warning system should be global, one of the most important components of any future network will be the national warning centres. Japan, Chile, New Zealand, Australia, French Polynesia, United States and the Russian Federation already run operational tsunami warning centres 24 hours a day, seven days a week. The track record of these centres is substantial, but it is time to improve the scope of their activities by working to build an operational, global ocean-disaster warning, preparedness and mitigation system.

> "The best way to ensure that a tsunami warning system remains fully operational for decades to come is to embed it in broader efforts to observe the ocean."

In addition to detecting multiple hazards—from storm surges to cyclones—the best way to ensure that a tsunami warning system remains fully operational for decades to come is to embed it in broader efforts to observe the ocean. Data used for tsunami warnings are of potential interest to an enormous array of users and stakeholders. It is these other users who will ensure the system is maintained over the long term.

For example, changes in observed sea level occur across many timescales, from seconds and minutes (wind waves, tsunamis), hours to days (tides, storm surges), and years (seasonal cycles, El Niño), through to long-term changes associated with climate change and the movement of land masses. Ocean circulation and long-term sea-level trends are monitored by the global array of tide gauges maintained by the Global Sea Level Observing System (Fig. 1 omitted) a component of the Global Ocean Observing System (GOOS). These are both run by the IOC, which aims to build a network of roughly 300 sea-level stations around the world (100 more than there are now), as well as several higher density regional networks.

Although some GLOSS stations already glean and process data in real time for the Pacific Tsunami Warning System, they operate mainly to serve the research community in a delayed mode. Upgrading the GLOSS network to real-time data delivery would contribute to a global tsunami warning system, and at the same time vastly increase its usefulness for other purposes.

For example, real-time sea-level data could contribute to ocean models serving a wide spectrum of users—including captains of large tankers who need predictions for efficient route planning. In such contexts, these data are of substantial economic interest. They can aid ship piloting in harbours, the management of sluices and barrages, tidal predictions and computations for coastal engineering design and insurance purposes.

The way forward

There are three substantial hurdles that need to be overcome to achieve this vision. The first challenge will be to develop an operational 'real time, all the time' capability for the ocean observing system. Those components of GOOS most relevant to marine hazards, such as sea surface temperature, and sea-level and seafloor pressure, need to be made available in real time. This is not just a technical requirement, but also a difficult political issue. For example, some countries purposely limit the release of public data to monthly mean sea-level values, years after the fact, whereas their high-frequency data (1-2 minute averages) are kept private for reasons ranging from cost to national security. In addition, national centres running operationally 24 hours a day, seven days a week, are essential to a hazard warning system. With the exception of a few countries, oceanography does not have the required institutional support at the national level to enable such operations, and creative solutions will be required.

The second challenge will be to bring together the different scientific communities, such as seismologists involved in tsunami warnings, meteorologists involved in storm-surge warnings, and oceanographers involved in both, to develop an integrated, multihazard system. So far, it has been difficult to build even single-use systems except at a national level. A fully operational multihaz-ard observing system will require unprecedented cooperation among a wide community of experts and stakeholders. But it would also dramatically improve cost-effectiveness, by both reducing the initial investment and spreading the burden of long-term costs.

The final and most difficult challenge will be to tailor the system to local cultural, social and economic conditions. Although the tsunami warning system must work on a global scale, its users will be local. As with so many things, we need to be thinking globally and acting locally. Civil populations cannot be educated or warned without accounting for—and benefiting from—local knowledge and concerns. Outreach, education and public awareness efforts will only work if they are woven into national, cultural and local environmental fabrics. For example, in Aceh, Indonesia, it has been suggested that rapid delivery of warnings could exploit the wide distribution of Islamic mosques with loudspeaker systems used for calls to prayer.

Ultimately, the development of the scientific and technical backbone of a tsunami warning system is a global responsibility, but preparedness remains a task for individual nations, or regions. This is the hardest of the three challenges and will require novel mechanisms for cooperation between scientists and social scientists, and between different organizations at the international, national and regional levels.

In particular, the international scientific community must not get carried away with the tantalizing but flawed idea that there is a quick technological fix to these complex societal issues. Instead, we need to broker a process through which countries of any given region come to recognize themselves as the true owners of the system. In their eagerness to help, states or organizations from outside the region might even obstruct the process by which Indian Ocean rim countries come together to plan, create and implement a system. But such a process should develop a true sense of ownership and responsibility. The majority of the lives lost were Asian, and the countries of that region must be at the forefront of plans to protect themselves in the future.

From 3 to 8 March 2005, UNESCO is hosting the first of two technical meetings intended to foster the development of a tsunami warning and mitigation system for the Indian Ocean. All of the nations in the region are invited and, along with other interested nations and international organizations, will work together to design a comprehensive work plan and timetable.

The challenge facing these countries, together with the IOC and our global partners, is a substantial one. But unlike so many visionary projects mooted by bureaucrats, the task is both clearly defined and eminently achievable. Let us hope that we are now taking the first step to ensure that the next tsunami—wherever and whenever it inevitably occurs—will not go down in history as a catastrophe, but as a tribute to the ability of science and technology to serve society.

Acknowledgments

I thank Thorkild Aarup, Bernardo Aliaga, Patricio Bernal, Ehrlich Desa, Albert Fischer, Paul Mason, Peter Pissiersens and Francois Schindele for contributing many useful thoughts to this article.

Keith Alverson is at the Global Ocean Observing System of the Intergovernmental Oceanographic Commission of UNESCO, 1 Rue Miollis, 75732 Paris, CEDEX 15, France.

AFTER APARTHEID ...

Change is everywhere in South Africa

Judith Fein

"He's our Moses," said a middle-aged black South African school teacher when my husband and I asked him how he felt about Nelson Mandela. "He led us out of the slavery of apartheid, through the Reed Sea (once called the Red Sea) of uncertainty and into the promised land of a democratic South Africa. It's only 10 years since apartheid officially ended. And you cannot imagine the change."

Change. It's exciting. It's vibrant. It's everywhere in South Africa, as people try to adapt and redefine themselves in relation to the new reality.

The black majority, despite huge problems like unemployment and AIDS, is walking easily through doors that were shut to them for decades. The white community—which had both supported and condemned apartheid after it was instituted in 1948—has been forced to do some soul searching and to accept the fact that it is no longer in charge.

In many cases today, members of the white community are last in line for plum jobs and government favors, and they are—excuse the expression—sweating through the changes.

In Johannesburg, beautiful, brainy, black tour operator Thuli Khumalo is multilingual (English, her native Zulu, German, Xhosa, Sotho and Tswana) and she whisks visitors through areas that either didn't exist or would have been off-limits a decade ago.

"Come, come, sisi (sister)," she said with her characteristic friendliness. "I am going to take you to our Apartheid Museum. It will teach you about what we went through."

Outside the entry, I sat for a moment on a bench and asked Thuli to join me.

"I can't, sisi," she said. "This bench is not for me."

I twisted around and saw a sign on the bench: "Europeans Only." Repulsed, I jumped up.

The entry to the museum was equally painful. Each of us, regardless of our color, was given a ticket stamped "white" or "non-white," and this determined which door we could use. My husband was white. I was nonwhite, separated from the whites. We were thrown into a simulation of the apartheid experience.

Inside the museum, there was little noise or chit-chat. Visitors walked down a hallway lined with the humiliating and hated passbooks that blacks had to carry for identity checks. Without them, they could be summarily arrested and jailed. Walls were lined with photos and documents about racial classification and bizarre laws governing who was deemed black or colored.

(A total of 2,823 people did the "chameleon dance" and had their color changed by the stroke of a government pen after they appealed their classification. In the year 1986 alone, 702 coloreds became white, 19 whites became colored, 249 blacks became colored and one Indian became white.)

In succeeding rooms, video screens played footage of Afrikaans nationalist propaganda, police beatings, black resis-

tance and interviews with Nelson Mandela, Winnie Mandela and murdered leader Steve Biko. Rooms were filled with jail cells, towering armored tanks and nooses used for lynchings. The corridors grew narrower, darker, the noise of screeching sirens increased, the videos multiplied, and the volume was cranked up until it became almost unbearable.

The small group of German tourists I was with exited the museum, saying they were exhausted. One man insisted he was overwhelmed. "That's what it was like for us, *boet* (brother)," said Thuli.

The next stop was at Gold Reef, one of the oddest pairings of tourism and tragedy I have experienced anywhere. Old gold mines provided wealth in Johannesburg but were the source of misery to the blacks, who were separated from their families and had to work there for a pittance.

Today, they have been turned into a Disneyland-like theme park. The old mine cars and equipment and the mine manager's house and administrative buildings have been transformed into rides and shops. Only when we were given hard hats and led into the bowels of the earth did we have a sense of the claustrophobic and brutal reality of the 28,000 mine workers.

The guide, a Chinese man deemed colored during apartheid, was upbeat and funny, and he made sure everyone had a grand time. He also confessed that guides were not allowed to give statistics about how many people died in the mines or other grisly details.

Anyone who remembers apartheid recalls Soweto (it means South West township) and the violence that erupted there as blacks rose to resist their oppressors.

Today, visitors are welcomed to Soweto, and Thuli drove us past the house where her grandparents lived, took us to the central market (which has everything from sandals made from tires to bowls crafted from telephone wires to beauty potions) and led us into Wandie's restaurant to sample delicious South African favorites like *pap* (corn porridge), *ting* (sorghum), *idombolo* (dumplings), *boerwoers* (beef sausages) and *gwinya* (fat cakes, an Afrikaans treat that has been fried in fat and has a sweet or savory filling).

After lunch, we rode by Mandela's house, his ex-wife Winnie's palatial residence and the square where 16-year-old Hector Peterson was shot by police in 1976 during a peaceful student demonstration. Then we visited a few of the houses and saw the shameful living conditions of some, the upgraded residences of others and the middle-class comfort of those who were upwardly mobile.

"Want to visit a *shebeen*?" a friendly Soweto boy asked us. "It's like a speak-easy." We followed him into his neighbor's small living room, which had been turned into a de facto bar. Music blared and unemployed locals hung out and drank beer, served ice cold by the woman of the house.

Back in the days of apartheid, *shebeens* served the same purpose—drinking, music and dancing—but were also meeting places to foment revolution. The police often raided the *shebeens* in a futile attempt to staunch those efforts.

One creative solution to unemployment is "unofficial" selling—along the principal streets and at busy intersections in Joburg. Artists and craftsmen ply their wares—and tourists can buy sculptures, jewelry, decorative items and paintings directly from the artists. Bargaining is de rigueur, and it is not unusual for vendors to settle for 30 percent, 40 percent or even 50 percent of their original asking price.

On weekends, the official Bruma flea market offers bargains on everything from clothes to art and souvenirs. The quality is uneven, but the prices are generally lower than the Rosebank Rooftop market, which is shoppers' heaven on Sundays. The spectacular handicrafts come from all over Africa.

For more adventurous art hounds, a three-hour trip from Joburg to the Ndebele tribe is well worth the visit. The women of the tribe paint their houses, inside and out, according to their inspirations. The brightly colored, geometrically patterned abodes are world famous.

Queen among the painters is Esther Mahlangu, who teaches the secrets of Ndebele art to children under a tree. Esther's art—painting and beadwork—is gobbled up by collectors, but if you are lucky, she will have pieces on hand to sell—and Thuli can act as translator because Esther's native Zulu is close to the Ndebele language.

If you can't buy from Esther, other Ndebele women sell beadwork and paintings in front of their houses at reasonable prices. The end of apartheid has opened up tribal villages to tourism for a much-needed infusion of capital.

In Cape Town, our guide, Tariek, drove past St. George's Cathedral, where Anglican Bishop Desmond Tutu had his congregation. During apartheid, guides said, the police hounded Tutu, arrested him and placed him under curfew, and his family was subjected to "humiliation." When apartheid was over, Tutu became the first black archbishop, and today he is revered as a hero.

Then Tariek took us to the area where he used to live and, growing teary-eyed, recounted how his home was razed and his family displaced in a forced removal—a fate 60,000 other black people in his district had to endure.

He was proud that he and the other coloreds took part in marches as part of their organized defiance. When I asked Tariek if he was angry, he said he didn't like to look backward. He was glad he could work—being a tour guide would have been impossible for a colored man during apartheid—and although the pace of economic growth wasn't always fast enough, it was, at last, happening.

"Black kids of today are not interested in apartheid. It's like our parents talking to us about World War II. … They never knew what it was like before. All our kids today live in a bubble, with no sense of the past."

JOHNNY CLEGG
rocker known as the "white Zulu" who was active in the anti-aparthied movement

---

The Waterfront area of Cape Town offers 380 shops, 72 restaurants and eateries, seven hotels and the opportunity to board a boat for the seven-mile pilgrimage to Robben Island, where Nelson Mandela was held prisoner for 18 years. Buses take visitors to the bleakest part of the island—torrid in summer and freezing in winter—

where the prisoners worked the limestone quarries with pick axes until the dust and the glare of the sun on the stone damaged their eyes. Then the tourists were deposited at the entry to the harsh prison, where their guide turned out to be a former political prisoner.

Our guide, Patrick Matanjana, spoke with acceptance and without bitterness about his incarceration, but admitted he took the job because he needed money, not because he wanted to come back to the locus of his suffering. He walked us through the washrooms, the yard and the small, sparse cell where Mandela was incarcerated. It felt like horrible but hallowed ground. It was here the politics of the new South Africa were forged by the modest, unassuming icon of the revolution.

Many people I met in Cape Town—blacks and whites alike—told me their memories of the day Nelson Mandela and other prisoners were released from Robben Island. The nation held its breath, because after a total of 27 years of incarceration, Mandela walked out a free man.

Many whites had fled the country (several black people I met referred to this as the "chicken run"), fearing a bloodbath. But in one of the miracles of the last century, the man whose spirit they tried to break came out of confinement and preached not revenge and violence but love and brotherhood. He became—and still is—a model for all humanity.

In Cape Town, a cab driver took us to Langa, another township. Unlike Soweto, the Langa we saw had no comfy homes with modern conveniences. It was subsistence living, with few amenities and crowded conditions. But Langa is using a lot of homegrown talent and ingenuity to compensate for economic hardship. It has a craft and gift shop, and locals run walking tours of the township.

"We had visitors coming through in cars, staring at us," said Shelly, our young guide. "We felt like they were on safari, and we were being stalked like animals. So we decided to do the tours ourselves."

Shelly spoke about her history, her family and what it was like growing up there. "The hardest thing about apartheid for me, personally, was that I never got to know my dad," she said. "You see, he was white, and so he wasn't allowed here. For this stupid reason, I could never see him, and he was a stranger to me. This has influenced all my life."

At the intimate and elegant Cape Grace Hotel, in Cape Town, black culture is now incorporated into the activities, the menu and even the spa treatments. The signature African Cape massage is based on the circular dance movements of the Khoi-San tribe. The massage medium includes shea butter, among other things.

In the Durban area, the end of apartheid meant native tribes could create employment by providing services to tourists. At Phe Zulu, visitors can experience Zulu traditions in a re-created village. I have heard that other villages are more touristy and found this one to be intimate, friendly, informative and rather delightful.

Young men in warrior garb perform vigorous dances and go through courting rituals with traditionally dressed young women.

A guide named Washington leads visitors into a grass hut and tells them how to make Zulu beer (from maize meal, sorghum malt and maize malt), what the colors in a beaded Zulu love letter mean (red means "my heart bleeds for you," blue means, "my love is endless like the sky" and black means "I'm as black as the rafters in the house from missing you") and about the *lobola*, or bride price, that a suitor must pay (11 cows).

At the highly recommended Phinda game reserve, democracy means blacks work alongside whites as managers, trackers, guides and security guards.

It is not only an awesome experience to find baby elephants frolicking in a lake, a pride of lions feeding on a freshly killed ze-

bra or a leopard darting through the bush at night, but it is also soul-stirring to sit around a campfire where a chef from Zimbabwe prepares food in a pit made from termite mounds mixed with termite saliva and hear tribal stories from the heart of Africa.

Before leaving South Africa, I spent some time with Johnny Clegg, the famed rocker who is referred to as the "white Zulu."

He was active in the antiapartheid movement, put his life on the line, militated for the miners and is considered a revolutionary brother by every black South African I spoke to.

What he said about young South Africans was arresting: "Black kids of today are not interested in apartheid. It's like our parents talking to us about World War II. They are aware of brands and the global youth culture. You tell them about the progress that has been made, but they can't relate. They never knew what it was like before. All our kids today live in a bubble, with no sense of the past. They are also worried about AIDS, drugs, the new branded products they are bombarded with. Apartheid isn't on their radar."

It certainly was on the radar of all the grown-up South Africans I met, and it was on my mind every day I spent in the country.

The movement from oppression to freedom has created a society that is vigorous, exciting and full of passion, surprises and opportunity. The emotions of the people

If you go ...

For general South Africa tourism information: www.southafrica.net
• South Africa Airways, which offers nonstop flights and has many special deals: www.flysaa.com
• The Cape Grace in Cape Town, one of the world's best boutique hotels: www.capegrace.com
• Phinda, part of the Conservation Corporation Africa group, which operates 30 camps and 15 lodges throughout Africa: www.ccafrica.com
• Wandie's restaurant in Soweto: www.guides.com/detail.cfm?detailID=187649
• Apartheid Museum: www.apartheidmuseum.org
• Gold Reef: goldreefcity.co.za
• Robben Island: www.freedom.co.za
• Phe Zulu: info@tourism-kzn.or
• Thuli Khumalo, a Zulu tour operator who arranges customized visits: ATAMELA@webmail.co.za

who have lived through apartheid and are willing to speak about it make a visit there unlike a visit to any other country: Go before it becomes a story in the history books.

From the *Santa Fe New Mexican*, June 13, 2004, pp. G1, G2. Copyright © 2004 by Judith Fein. Reprinted by permission of the author.

UNIT 2

Human-Environment Relations

Unit Selections

Key Points to Consider

- What are the long-range implications of atmospheric pollution? Explain the greenhouse effect.

- How can the problem of regional transfer of pollutants be solved?

- The manufacture of goods needed by humans produces pollutants that degrade the environment. How can this dilemma be solved? Does the Indonesian logging plan have the answer?

- Where in the world are there serious problems of desertification and drought? Why are these areas increasing in size?

- How are you, as an individual, related to the land? Does urban sprawl concern you? Explain.

- In your view, is global warming an accepted fact?

Student Website

www.mhcls.com/online

Internet References

Further information regarding these websites may be found in this book's preface or online.

Alliance for Global Sustainability (AGS)
http://www.global-sustainability.org

Human Geography
http://www.geog.le.ac.uk/cti/hum.html

The North-South Institute
http://www.nsi-ins.ca

United Nations Environment Programme (UNEP)
http://www.unep.ch

US Global Change Research Program
http://www.usgcrp.gov

World Health Organization
http://www.who.int

The home of humankind is Earth's surface and the thin layer of atmosphere enveloping it. Here the human populace has struggled over time to change the physical setting and to create the telltale signs of occupation. Humankind has greatly modified Earth's surface to suit its purposes. At the same time, we have been greatly influenced by the very environment that we have worked to change.

This basic relationship of humans and land is important in geography. In unit 1, William Pattison identified this relationship as one of the four traditions of geography. Geographers observe, study, and analyze the ways in which human occupants of Earth have interacted with the physical environment. This unit presents a number of articles that illustrate the theme of human-environment relationships. In some cases, the association of humans and the physical world has been mutually beneficial; in others, environmental degradation has been the result.

At the present time, the potential for major modifications of the Earth's surface and atmosphere is greater than at any other time in history. It is crucial that the environmental consequences of these modifications be clearly understood before such efforts are undertaken.

A plan to harvest logs in an Indonesian forest without degrading the environment is discussed in the first article. The next article summarizes the negative effects of global warming. "Environmental Enemy No. 1" reviews ways to reduce carbon dioxide accumulations in the atmosphere. "A Great Wall of Waste" draws attention to the enormity of environmental pollution as China industrializes. The next article deals with the growth of coal-fired power plants in small midwestern towns.

This unit provides a small sample of the many ways in which humans interact with the environment. The outcomes of these interactions may be positive or negative. They may enhance the position of humankind and protect the environment, or they may do just the opposite. We human beings are the guardians of the physical world. We have it in our power to protect, to neglect, or to destroy the physical world.

The Race To Save A Rainforest

Can an experiment in Indonesia prove the merits of sustainable logging to Big Timber?

T HE INDONESIAN VILLAGE of Long Pai lies a grueling three-hour drive on rutted dirt roads from the provincial capital of Tanjung Redeb. It is a river town in Borneo that is little changed from a century ago when it served as the backdrop for Joseph Conrad's *Lord Jim*. Agus Heryanto, a local representative of the Nature Conservancy, a Virginia-based environmental group, has made the bone-rattling trip to convince villagers that they should embrace a plan to log the lush forests nearby. By the dim light of a single bulb in a stilt hut, Agus tells village elders about a trust fund supported by logging revenues that will pay for roads, medical clinics, and electric generators. And, promises Agus, the logging will be done in an environmentally responsible way.

The villagers listen to the pitch, but remain suspicious. Three years earlier, they drove out the company that wants to do the logging—Sumalindo Lestari Jaya—after disputes over payments and improper cutting of valuable trees. Agus says that this time the trust will protect against wrongdoing. Finally, at a second meeting four days later, the Long Pai elders agree to the plan. By yearend they hope to have a formal contract.

There seems to be a disconnect here. Indonesia is known for the rampant destruction of its rainforest. Now one of the country's most powerful logging companies allies itself with a Western environmental group to make nice with impoverished villagers? What happened in that stilt hut is an exampe of globalization at work—an intricate dance involving Western consumers, multinational companies, environmental activists, na-

tive forest dwellers, and local conglomerates. The aim is to combine the profit motive with the need to end the shocking degradation of the forest. A long shot perhaps, but some action is desperately needed, and it is starting in Borneo.

REWARD SYSTEM

THE SUMALINDO DEAL was indirectly driven by the vigorous campaigning of environmental groups. They have put pressure on Western retailers such as Home Depot (HD), Ikea, and Kinko's to stop buying wood, pulp, and paper from companies that do not follow sustainable practices. The retailers in turn are leery of infuriating shoppers appalled by the rape of the forests. "We can give logging companies an incentive to manage the forest," says Nigel Sizer, director of the Asia-Pacific Forests program at the Nature Conservancy. The effort is already showing results: Sumalindo, afraid of losing access to the all-important U.S. and European markets, is working with the Nature Conservancy to clean up the way it cuts timber.

The initiative still has a long way to go. Sumalindo is one of only a handful of enterprises taking even modest steps toward low-impact logging. Still, if the Indonesian effort is successful, it could serve as a template for saving forests in Southeast Asia, the Amazon basin, and Siberia.

BAR CODES VS. BOYCOTTS

BORNEO'S FORESTS don't have much time. Many of the island's great stands of trees have been felled, and Indonesian

loggers continue to decimate an area half the size of Switzerland each year. Environmentalists warn that Borneo's wild orangutans, sun bears, and clouded leopards could be wiped out in 10 to 20 years, as well as countless other species. Plus, thousands of loggers could end up jobless in a land so scarred that it may be nearly impossible to revive it for other commercial uses.

Environmentalists hope to drive many more efforts like Sumalindo's. In a project set to begin before yearend, the Nature Conservancy has teamed up with Home Depot Inc. and a British aid agency to track lumber from the stump to the store. Logs and boards will be tagged with bar codes that allow buyers to determine whether their wood was harvested in a sustainable manner. "People were asking what we were doing not to add to the deforestation of the world," says Ron Jarvis, head of lumber merchandising at Home Depot. "We didn't have the answer."

Where Home Depot and the Nature Conservancy are offering carrots, more radical environmentalists are wielding sticks. Rainforest Action Network and Greenpeace International want consumers to stop buying Indonesian timber products. They think the scale of illegal Indonesian logging is so vast and the legal concessions so destructive that reforms like those proposed by the Nature Conservancy are doomed. Michael Brune, executive director of San Francisco-based Rainforest Action, says his group has pushed for a boycott of Indonesian timber products for two years. "This is not a time to establish small toeholds of good production."

The protests are having an effect. Home Depot has cut its purchases of lumber from Indonesia by more than three-quarters since 2000. In early 2002, the American Forest & Paper Assn. trade group said its members would stop buying illegal wood products, singling out Indonesia as a special concern. And Ikea requires that all of its tropical hardwoods be cut in areas okayed by the Forest Stewardship Council, a German group that issues certificates to companies following sustainable practices. Ikea cut off Indonesian teak purchases in October, 2001, when battles between local villagers and a government-owned supplier in Java over ownership of the trees short-circuited the certification process. "We regret that it happened," says Ikea forestry coordinator Par Stenmark. "But it shows that the Forest Stewardship Council certification system is working."

Pressure from Western shoppers is felt in the forest

A boycott is just the outcome Sumalindo wants to avoid—because the corporate damage could go far beyond Sumalindo itself. The company is owned by one of Indonesia's largest wood empires, the Hasko Group.

Consequently, Sumalindo is making an effort to change. In a timber concession 200 kilometers south of Long Pai, it has worked with German and U.S. experts to minimize damage to the forest and is aiming for certification from the Forest Stewardship Council. That's a sacrifice, given that the three-week evaluation costs some $50,000. Other costs may ultimately prove much higher: Sumalindo had to agree not to cut some 212,000 acres—nearly a third of one of its concessions.

Sumalindo is also paying for a variety of measures to lessen its environmental impact. These include a $500,000 aerial tram that lifts logs out of steep valleys rather than pulling them through the forest, which can cause erosion. Nature Conservancy experts also teach loggers how to fell a tree so that it ends up closer to a road, minimizing damage to the forest floor by reducing dragging. And they counsel against slashing through the forest's thick underbrush with a bulldozer.

SLASHERS AND POACHERS

THE DIFFERENCE between Sumalindo's methods and those of other companies—especially those that are government-owned—is stark. Just beyond the borders of Sumalindo's land in one of Borneo's most important watersheds lies a concession owned by the government logging company, PT Inhutani. Much of the acreage has already been cut. Gulleys full of eroded tropical soil abut the road. Wildfires raged across huge swaths of the land five years ago, after intensive logging. The company declined to comment.

Even worse are the poachers. These illegal loggers cut with impunity, often working in plain sight along main roads. They can take down a 100-year-old tree in under 10 minutes. On the banks of the Segah River, which runs through the area, some 300 logs waiting for shipment lacked any identifying marks—a telltale sign of illicit cutting. Most of this wood is smuggled across the border to Malaysia, where it can be imported with a small bribe, according to Indonesian officials and environmental groups. A senior Malaysian official, saying his country takes smuggling "very seriously," notes that Malaysia has banned imports of some wood from Indonesia. Much, though, still gets through, and eventually finds its way to mills in China. "Illegal logging is our responsibility," says Wahjudi Wardojo, the highest ranking civil servant at Indonesia's Forestry Ministry. "But without support from other countries, it's useless."

So illegal logging continues—and even most of the legal logging in Indonesia is not done sustainably. If contracts such as the proposed pact between Sumalindo and Long Pai become the norm, villagers will have a bottom-line incentive to see that all logs that are cut are accounted for, and the tracking system should ensure that the timber can be traced. But if other companies—and countries—don't sign on, the effort is doomed. And it remains to be seen if Western consumers will pay a premium for wood harvested sustainably. "We are struggling," says Wahjudi. "Our time is very limited." If they don't succeed, another corner of the earth's forests will be gone forever.

By Mark L. Clifford in Long Pai, Indonesia, with Hiroko Tashiro in Tokyo and Anand Natarajan in Atlanta

---

GLOBAL WARMING

Consensus is growing among scientists, governments, and business that they must act fast to combat climate change. This has already sparked efforts to limit CO_2 emissions. Many companies are now preparing for a carbon-constrained world.

JOHN CAREY

THE IDEA THAT THE HUMAN species could alter something as huge and complex as the earth's climate was once the subject of an esoteric scientific debate. But now even attorneys general more used to battling corporate malfeasance are taking up the cause. On July 21, New York Attorney General Eliot Spitzer and lawyers from seven other states sued the nation's largest utility companies, demanding that they reduce emissions of the gases thought to be warming the earth. Warns Spitzer: "Global warming threatens our health, our economy, our natural resources, and our children's future. It is clear we must act."

The maneuvers of eight mostly Democratic AGs could be seen as a political attack. But their suit is only one tiny trumpet note in a growing bipartisan call to arms. "The facts are there," says Senator John McCain (R-Ariz.). "We have to educate our fellow citizens about climate change and the danger it poses to the world." In January, the European Union will impose mandatory caps on carbon dioxide and other gases that act like a greenhouse over the earth, and will begin a market-based system for buying and selling the right to emit carbon. By the end of the year, Russia may ratify the Kyoto Protocol, which makes CO_2 reductions mandatory among the 124 countries that have already accepted the accord. Some countries are leaping even further ahead. Britain has vowed to slash emissions by 60% by 2050. Climate change is a greater threat to the world than terrorism, argues Sir David King, chief science adviser to Prime Minister Tony Blair: "Delaying action for a decade, or even just years, is not a serious option."

There are naysayers. The Bush Administration flatly rejects Kyoto and mandatory curbs, arguing that such steps will cripple the economy. Better to develop new low-carbon technologies to solve problems if and when they appear, says Energy Secretary Spencer Abraham. And a small group of scientists still argues there is no danger. "We know how much the planet is going to warm," says the Cato Institute's Patrick J. Michaels. "It is a small amount, and we can't do anything about it."

But the growing consensus among scientists and governments is that we can—and must—do something. Researchers under the auspices of the National Academy of Sciences and the Intergovernmental Panel on Climate Change (IPCC) have pondered the evidence and concluded that the earth is warming, that humans are probably the cause, and that the threat is real enough to warrant an immediate response. "There is no dispute that the temperature will rise. It will," says Donald Kennedy, editor-in-chief of *Science*. "The disagreement is how much." Indeed, "there is a real potential for sudden and perhaps catastrophic change," says Eileen Claussen, president of the Pew Center on Global Climate Change: "The fact that we are uncertain may actually be a reason to act sooner rather than later."

Plus, taking action brings a host of ancillary benefits. The main way to cut greenhouse-gas emissions is simply to burn less fossil fuel. Making cars and factories more energy-efficient and using alternative sources would make America less dependent on the Persian Gulf and sources of other imported oil. It would mean less pollution. And many companies that have cut emissions have discovered, often to their surprise, that it saves money and spurs development of innovative technologies. "It's impossible to find a company that has acted and has not found benefits," says Michael Northrop, co-creator of the Climate Group, a coalition of companies and governments set up to share such success stories.

That's why there has been a rush to fill the leadership vacuum left by Washington. "States have stepped up to fill this policy void, as much out of economic self-interest as fear of devastating climate changes," says Kenneth A. Colburn, executive director of Northeast States for Coordinated Air Use Management. Warning of flooded coasts and crippled industries, Massachusetts unveiled a plan in May to cut emissions by 10% by 2020. In June, California proposed 30% cuts in car emissions by 2015. Many other states are weighing similar actions.

Curbing Carbon

REMARKABLY, BUSINESS is far ahead of Congress and the White House. Some CEOs are already calling for once-unthinkable steps. "We accept that the science on global warming is overwhelming," says John W. Rowe, chairman and CEO of Exelon Corp. "There should be mandatory carbon constraints."

Exelon, of course, would likely benefit as the nation's largest operator of commercial nuclear power plants. But many other companies also are planning for that future. American Electric Power Co. once fought the idea of combating climate change. But in the late 1990s, then-CEO E. Linn Draper Jr. pushed for a strategy shift at the No. 1 coal-burning utility—preparing for limits instead of denying that global warming existed. It was a tough sell to management. Limits on carbon emissions threaten the whole idea of burning coal. But Draper prevailed. Why? "We felt it was inevitable that we were going to live in a carbon-constrained world," says Dale E. Heydlauff, AEP's senior vice-president for environmental affairs.

Now, AEP is trying to accumulate credits for cutting CO_2. It's investing in renewable energy projects in Chile, retrofitting school buildings in Bulgaria for greater efficiency, and exploring ways to burn coal more cleanly. Scores of other companies are also taking action—and seeing big benefits. DuPont has cut its greenhouse-gas emissions by 65% since 1990, saving hundreds of millions of dollars in the process. Alcoa Inc. is aiming at a 25% cut by 2010. General Electric Co. is anticipating growing markets for its wind power division and for more energy-efficient appliances. And General Motors Corp. is spending millions to develop hydrogen-powered cars that don't emit CO_2. A low-carbon economy "could really change our industry," says Fred Sciance, manager of GM's global climate issues team. As Exelon knows, the need for carbon-free power could even mean a boost for advanced nuclear reactors, which produce electricity without any greenhouse-gas emissions.

Global warming could change other industries, too. Even if the world manages to make big cuts in emissions soon, the earth will still warm several more degrees in coming decades, most climate scientists believe. That could slash agricultural yields, raise sea levels, and bring more extreme weather.

For businesses, this presents threats—and opportunities. Insurers may face more floods, storms, and other disasters. Farm-

ers must adjust crops to changing climates. Companies that pioneer low-emission cars, clean coal-burning technology, and hardier crop plants—or find cheap ways to slash emissions—will take over from those that can't move as fast. "There is no silver bullet," says Chris Mottershead, distinguished adviser at BP PLC: "There is a suite of technologies that are required, and we need to unleash the talent inside business" to develop them.

Are we ready for this carbon-constrained, warming world? In some ways, yes. "There is a case to be made for cautious optimism, that we are making small steps," says BP's Mottershead.

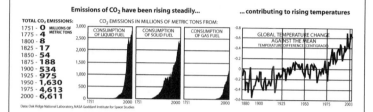

MANY SCIENTISTS AGREE ON THE BASICS OF GLOBAL WARMING ...

... AND THE EFFECTS ON THE PLANET COULD BE DIRE

1. FLOODING Seawaters could rise almost a meter in this century, and continue going up. Some coastal regions already see seasonal flooding, and the situation would get worse as water levels rise.

2. OCEAN DISRUPTIONS Coral reefs are under pressure from changes in water level and temperature. As more carbon goes into the sea, plankton could suffer, and that would affect species higher up the food chain.

3. SHIFTING STORM PATTERNS There are no data to show an increase in violent storms right now, but many scientists believe warming will bring more violent and unpredictable climate events.

4. REDUCED FARM OUTPUT In certain regions, each degree rise in the surface temperature brings a further drop in crop yields.

5. ANIMAL EXTINCTIONS Some species are already moving to cooler regions—and some aren't making it. Global warming may not yet be a factor, but it will almost certainly take its toll on species.

6. DROUGHTS In past periods of climate change, whole sections of Africa turned to desert. In extreme scenarios, areas that are currently fertile could become barren and dry.

Indeed, there is surprising consensus about the policies needed to spur innovation and fight global warming. The basic idea: mandatory reductions or taxes on carbon emissions, combined with a worldwide emissions-trading program. Here's how it could work: Imagine that each company in a particular sector is required to cut emissions by 20%. The company could meet the target on its own by becoming more energy efficient or by switching from fossil fuels to alternatives. But it could also simply buy the needed reductions on the open market from others who have already cut emissions more than required, and who thus have excess emissions to sell. Under a sophisticated worldwide carbon-trading system, governments and companies could also get sellable credits for planting trees to soak up carbon or for investing in, say, energy efficient and low-carbon technologies in the developing world. As a result, there is a powerful incentive for everyone to find the lowest-cost and most effective cuts—and to move to lower-carbon technologies.

A key element is long-term predictability. If the world sets goals for the next 50 years, as Britain has done, and then implements the curbs or taxes needed to reach them, companies will figure out solutions. "Give us a date, tell us how much we need to cut, give us the flexibility to meet the goals, and we'll get it done," says Wayne H. Brunetti, CEO and chairman of Xcel Energy Inc., the nation's fourth-largest electricity and gas utility.

The Challenge

SUCH CLEAR POLICY SIGNALS should bring major efficiency gains. Even 30% to 40% reductions in emissions by 2020 are possible, says Northrop. After that, he suggests, shifts to new energy technologies "can get the other 35% to 40% that we need to get to the low-carbon emission future."

The good news is that the world sees the threat and has begun to respond. The bad news is the magnitude of the task. Rising CO_2 levels in the atmosphere can't be slowed or reduced if only a few countries—or even all the industrialized nations—take action. The world must also figure out a way to permit growth in China, India, and other developing nations while lowering consumption of coal, gasoline, and other fossil fuels. "It's hard to think of a public policy issue that is harder than this one," says economist Jeffrey D. Sachs, director of Columbia University's Earth Institute.

Developing countries are responsible for just over one-third of the world's greenhouse-gas emissions. But they emit less than one-fifth as much per person as do the industrialized nations. That will increase as their citizens buy more cars and consume more energy. By 2100, these countries will emit two or three times as much as the developed world, experts predict.

The Bush Administration and Congress have seized upon this issue as one reason for rejecting the Kyoto Protocol, which doesn't include the developing world. But international negotiators are beginning to talk about a plan that would go beyond Kyoto. The first step: showing that the industrialized world is serious about leading the way. That's one of the motivations behind Britain's vow to slash emissions by 60%, for example. Britain knows it can't solve this global problem by itself. But committing to reducing CO_2 "is the right thing to do," says Brit-ish Energy Minister Stephen Timms. It will also keep the country from becoming dependent on foreign oil when its North Sea oil fields start to run dry in a few years.

The next step is to help the developing world adopt new technologies. China and other nations could avoid the West's era of gas-guzzlers and dirty power plants by jumping to highly efficient clean coal plants and hybrid or advanced diesel cars. What's needed, experts say, are incentives to stimulate companies to make investments in advanced technology in developing countries. Once an international carbon-trading system is put in place, suggests Elliot Diringer, director of international strategies at the Pew Center on Global Climate Change, "we can reduce our own costs in the U.S. by allowing our companies to get the benefit of low-cost emissions abroad."

The past few decades are the warmest since people began keeping records— changing the planet's face

Still, even if the developing world comes on board, staggering reductions in emissions are needed. Consider the math. For the past 450,000 years, the amount of carbon dioxide in the atmosphere has stayed below 290 parts per million (ppm). Now, we are spewing out more than 7 gigatons of carbon a year and large amounts of other greenhouse gases such as methane. As a result, the CO_2 levels in the air have climbed past 370 ppm. With no action, those levels could jump to 800 to 1,000 ppm by the end of the century. "We are already in dire straits," warns Columbia University geophysicist Klaus S. Lackner.

The Science

CAN SERIOUS CONSEQUENCES be prevented? The British government, many scientists, and some executives are urging an all-out effort to keep the earth from warming more than two degrees Celsius. "The consequences of changes above two degrees are so dreadful that we need to avoid it," says BP's Mottershead. To hit that target, scientists calculate that CO_2 concentrations in the atmosphere must be kept from reaching 550 ppm—twice the preindustrial level. Getting there may require cutting the world's per capita emissions in half by 2100.

Of course, there is great uncertainty surrounding the science of global warming. No one can really know the size and consequences of climate change. "Without a doubt, it will be a very different world—a much warmer world," says David S. Battisti, atmospheric scientist at the University of Washington. But how much warmer? Which regions will be better or worse off? Will there be more floods and droughts? There's even a chance of surprises beyond the scary predictions of some computer models. "What's worrisome are the unknown unknowns," says Daniel P. Schrag, director of the Laboratory for Geochemical Oceanography at Harvard University. "We are performing an experiment that hasn't been done in millions of years, and no one knows exactly what's going to happen."

What scientists do know is that carbon dioxide and a number of other gases act like the roof of a greenhouse. Energy from the sun passes through easily. Some of the warmth that normally would be radiated back out to space is trapped, however, warming the planet. With no greenhouse gases at all in the atmosphere, we would freeze. The earth's average temperature would be a cold -17C, not the relatively balmy 14C it is today.

But the atmosphere is fiendishly complicated. If an increase in greenhouse gases also makes the sky cloudier, the added clouds may cool the surface enough to offset warming from CO_2. Tiny particles from pollution also exert warming or cooling effects, depending on where they are in the atmosphere. Naysayers argue that it's just too soon to tell if greenhouse gases will significantly change the climate.

Yet the climate is changing. In the past 100 years, global temperatures are up 0.6 degrees Celsius. The past few decades are the warmest since people began keeping temperature records—altering the face of the planet. For instance, the Qori Kalis glacier in Peru is shrinking at a rate of 200 meters per year, 40 times as fast as in 1978. It's just one of hundreds of glaciers that are vanishing. Ice is disappearing from the Arctic Ocean and Greenland. More than a hundred species of animals have been spotted moving to cooler regions, and spring starts sooner for more than 200 others. "It's increasingly clear that even the modest warming today is having large effects on ecosystems," says ecologist Christopher B. Field of the Carnegie Institution. "The most compelling impact is the 10% decreasing yield of corn in the Midwest per degree [of warming.]"

More worrisome, scientists have learned from the past that seemingly small perturbations can cause the climate to swing rapidly and dramatically. Data from ice cores taken from Greenland and elsewhere reveal that parts of the planet cooled by 10 degrees Celsius in just a few decades about 12,700 years ago. Five thousand years ago, the Sahara region of Africa was transformed from a verdant lake-studded landscape like Minnesota's to barren desert in just a few hundred years. The initial push—a change in the earth's orbit—was small and very gradual, says geochemist Peter B. deMenocal of Columbia University's Lamont-Doherty Earth Observatory. "But the climate response was very abrupt—like flipping a switch."

The earth's history is full of such abrupt climate changes. Now many scientists fear that the current buildup of greenhouse gases could also flip a global switch. "To take a chance and say these abrupt changes won't occur in the future is sheer madness," says Wallace S. Broecker, earth scientist at Lamont-Doherty. "That's why it is absolutely foolhardy to let CO_2 go up to 600 or 800 ppm."

Indeed, Broecker has helped pinpoint one switch involving ocean currents that circulate heat and cold. If this so-called conveyor shuts down, the Gulf Stream stops bringing heat to Europe and the U.S. Northeast. This is not speculation. It has happened in the past, most recently 8,200 years ago.

Can it happen again? Maybe. A recent Pentagon report tells of a "plausible ... though not the most likely" scenario, in which the conveyor shuts off. "Such abrupt climate change ... could potentially destabilize the geopolitical environment, leading to skirmishes, battles, and even war," it warns.

There are already worrisome signs. The global conveyor is driven by cold, salty water in the Arctic, which sinks to the bottom and flows south. If the water isn't salty enough—thus heavy enough—to sink, the conveyor shuts down. Now, scientists are discovering that Arctic and North Atlantic waters are becoming fresher because of increased precipitation and melting. "Over the past four decades, the subpolar North Atlantic has become dramatically less salty, while the tropical oceans have become saltier," observed William B. Curry of the Woods Hole Oceanographic Institution in recent congressional testimony. "These salinity changes are unprecedented in the relatively short history of the science of oceanography."

If the global switch does flip, an Ice Age won't descend upon Europe, scientists now believe. But that doesn't mean the consequences won't be severe. The sobering lesson from the past is that the climate is a temperamental beast. And now, with the atmosphere filling with greenhouse gases, "the future may have big surprises in store," says Harvard's Schrag.

In some scenarios, the ice on Greenland eventually melts, causing sea levels to rise 18 feet. Melt just the West Antarctic ice sheet as well, and sea levels jump another 18 feet. Currently shrinking glaciers may mean threats to water supplies for farmers and cities. Meanwhile, higher temperatures can cut crop yields, inhibit rice germination, and devastate biologically vital ecosystems like coral reefs. A paper in the July 16 issue of Science suggests that increasing CO_2 levels in the ocean could affect the growth of marine life, with consequences for the oceanic food chain.

Prevent or Adapt?

PERHAPS THE CENTRAL debate in global warming now is not about the underlying science, but whether it's better—and cheaper—to take steps to prepare for or prevent climate change now, or to simply roll with the punches if and when it happens. Opponents of greenhouse-gas curbs say we should be able to adapt to a warmer world or even cool it back down. "I'm convinced there will be engineering schemes that will allow our children's children to have whatever climate they want," says Robert C. Balling Jr., a climatologist at Arizona State University and coauthor of *The Satanic Gases*, which argues that the worries are vastly overblown.

Yes, human beings can adapt, advocates of immediate action retort. But why run even the small risk of catastrophic changes, when important steps can be taken at a modest cost now? A British government panel, for instance, concluded that the cost of its share of the task of limiting the level of CO_2 to 550 ppm would be about 1% of Britain's gross domestic product.

By taking action to thwart global warming, companies may not only cut costs but also spark tech innovation

Compare that, says Sir David King, with the cost of a single flood breaking through the barrier in the Thames River—some 30 billion pounds, or 2% of current GDP. "Common sense says that it's time to purchase some low-cost insurance now," says economist Paul R. Portney, president of Resources for the Future.

The Business Response

WHEN CEOS CONTEMPLATE global warming, they see something they dread: uncertainty. There's uncertainty about what regulations they will have to meet and about how much the climate will change—and uncertainty itself poses challenges. Insurance giant Swiss Re sees a threat to its entire industry. The reason: Insurers know how to write policies for every conceivable hazard based on exhaustive study of the past. If floods typically occur in a city every 20 years or so, then it's a good bet the trend will continue into the future. Global warming throws all that historical data out the window. One of the predicted consequences of higher greenhouse-gas levels, for instance, is more variable weather. Even a heat wave like the one that gripped Britain in 1995 led to losses of 1.5 billion pounds, Swiss Re calculates. So an increase in droughts, floods, and other events "could be financially devastating," says Christopher Walker, a Swiss Re greenhouse-gas expert.

That's why Swiss Re has been pressing companies to plan for possible effects of warming. Lenders may require beefed-up flood insurance before issuing mortgages. Chipmakers must find replacements for greenhouse-gas solvents. Utilities need to prepare grids to handle bigger loads and to boost power from renewable sources. Oil companies need to think about a future where cars use less gas—or switch to hydrogen.

Swiss Re says the word is getting out, but not fast enough. In a recent survey, "80% of CEOs said that climate change was a potential risk, but only 40% were doing something about it," says Walker. "That's not good to hear for insurers."

Shareholders are also demanding that companies assess the risks of global warming and devise coping strategies. Moreover, multinationals have no choice but to plan for emissions cuts because of the coming EU carbon limits and possible restrictions on other greenhouse gases.

Intel Corp., for example, is worried the EU could ban the use of perfluorocarbons (PCF), chemicals used in chipmaking that are potent greenhouse gases. "We are looking for substitutes but don't have any yet," says Intel's Stephen Harper. "We decided to craft a worldwide agreement to reduce PFC emissions 10% by 2010—upwards of a 90% reduction per chip. We wanted to show leadership and not have the EU regulate us."

Utilities face the greatest threat since the bulk of the power they generate comes from climate-changing fossil fuels. That's why AEP, Cinergy Corp., and others are probing new technologies that would enable them to capture the carbon as coal is burned. That carbon could then be pumped deep into the ground to be stored for thousands of years. AEP has helped drill a test well in West Virginia to see if this sort of "carbon sequestration" is feasible and safe. And dozens of utilities are turning to alternative fuels, from wind to biomass. Florida Power & Light Co. now has 42 wind power facilities and has pushed energy efficiency, reducing emissions and eliminating the need to build 10 midsize power plants, according to Randall R. LaBauve, vice-president for environmental services. "We are seeing more companies committed to voluntary or even mandatory reductions," he says. Renewable energy, not counting hydropower, now produces only 2% of the nation's electricity. But some

states—along with Presidential candidate John Kerry—are proposing that this be increased to as high as 20%.

Who Will Lead?

EVEN WITHOUT MANDATES, scores of companies are taking concrete actions. "The science debate goes on, but we know enough to move now," explains AEP Chief Executive Michael G. Morris. It helps that thwarting global warming often brings cost savings and business benefits. Indeed, one goal of the newly formed Climate Group is to share tales of how climate strategies helped the bottom line. "The ones who have been at it for a while are finding they can do more than is asked for in Kyoto, and are achieving all kinds of benefits," says Northrop. BP, for instance, developed its own internal strategy for trading carbon emissions. That prompted a companywide search to find the lowest-cost reductions. Many of the measures were simple, such as identifying and plugging leaks. The overall result: a 10% reduction in emissions and a $650 million boost to the company in three years.

Climate-savvy execs are hoping that when carbon limits are imposed, they'll get credit for actions already taken. But they're also anticipating big future opportunities. GE bought Enron Corp.'s wind business and a solar energy company in addition to doing research on hydrogen and lower-emission jet engines and locomotives. "We can help our customers meet the challenges they are going to face," says Stephen D. Ramsey, GE's environmental chief. In Arizona, startup Global Research Technologies LLC is developing systems that use solvents to grab CO_2 out of the air and isolate it for disposal.

Given this progress, many scientists wonder why the world—and especially the U.S.—isn't moving faster to reduce the chances that global warming will bring nasty surprises. The reason for the inaction is "not the science and not the economics," says G. Michael Purdy, director of Lamont-Doherty. "Rather it is the lack of public knowledge, the lack of leadership, and the lack of political will."

The Bush Administration counters that taking steps is simply too costly. Imposing limits on the U.S. would throttle growth and put America at a competitive disadvantage around the world. "No nation will mortgage its growth and prosperity to cut greenhouse-gas emissions," says Energy Secretary Abraham. In any case, the White House is not ignoring the issue. It has called for voluntary reductions and it is funding research into new technologies. "If we are successful in developing carbon sequestration and cars that run on hydrogen fuel cells, that solves most of the problem with global warming," Abraham argues. "We may disagree on targets, but no one is going to reach any targets if we don't make these investments."

But most experts believe that mandatory curbs are essential and that they can be implemented at reasonable cost. Indeed, as states jump in with their own patchwork of rules, execs are beginning to say that it may be time to push for uniform national limits. That's what happened in 1990 with pollution rules. Faced with the prospect of dozens of state regulations, companies helped push for federal Clean Air Act amendments that reduced sulfur dioxide emissions through a market-based trading system. The law was a huge success. "We reduced emissions ahead of schedule and at lower cost," says Xcel Energy CEO Brunetti. "It's a great example of what can be done."

The same sort of trading scheme would bring similarly inexpensive greenhouse-gas reductions, many economists, politicians, and execs believe. The EU plan puts a cap on emissions for each country and allows emitters to buy and sell permits to release certain amounts of emissions. In the U.S., a market for trading carbon emissions—the Chicago Climate Exchange—already operates. And a bill to set up a cap-and-trade scheme, introduced by Senators John McCain and Joseph I. Lieberman (D-Conn.), is expected to win more votes than the 43 it garnered—against the odds—last year.

These steps are just the beginning, though. Even drastic measures—such as implementing revolutionary energy technologies or grabbing carbon from the air—won't stop this great global experiment from being conducted. "We won't cure this problem," cautions Henry Jacoby, co-director of Massachusetts Institute of Technology's Joint Program on the Science & Policy of Global Change. "The hope is that we can lower the risk of some of the more possible damaging outcomes." Companies and nations have begun to respond, but there is a long way to go, and only two choices: Get serious about global warming—or be prepared for the consequences.

—*With Sarah R. Shapiro in New York*

Environmental enemy No. 1

Cleaning up the burning of coal would be the best way to make growth greener

IS GROWTH bad for the environment? It is certainly fashionable in some quarters to argue that trade and capitalism are choking the planet to death. Yet it is also nonsense. As our survey of the environment this week explains, there is little evidence to back up such alarmism. On the contrary, there is reason to believe not only that growth can be compatible with greenery, but that it often bolsters it.

This is not, however, to say that there are no environmental problems to worry about. In particular, the needlessly dirty, unhealthy and inefficient way in which we use energy is the biggest source of environmental fouling. That is why it makes sense to start a slow shift away from today's filthy use of fossil fuels towards a cleaner, low-carbon future.

There are three reasons for calling for such an energy revolution. First, a switch to cleaner energy would make tackling other green concerns a lot easier. That is because dealing with many of these—treating chemical waste, recycling aluminium or incinerating municipal rubbish, for instance—is in itself an energy-intensive task. The second reason is climate change. The most sensible way for governments to tackle this genuine (but long-term) problem is to send a powerful signal that the world must move towards a low-carbon future. That will spur all sorts of innovations in clean energy.

The third reason is the most pressing of all: human health. In poor countries, where inefficient power stations, sooty coal boilers and bad ventilation are the norm, air pollution is one of the leading preventable causes of death. It affects some of the rich world too. From Athens to Beijing, the impact of fine particles released by the combustion of fossil fuels, and especially coal, is among today's biggest public-health concerns.

Dethroning King Coal

The dream of cleaner energy will never be realised as long as the balance is tilted toward dirty technologies. For a start, governments must scrap perverse subsidies that actually encourage the consumption of fossil fuels. Some of these, such as cash given by Spain and Germany to the coal industry, are blatantly wrong-headed. Others are less obvious, but no less damaging. A clause in America's Clean Air Act exempts old coal plants from complying with current emissions rules, so much of America's electricity is now produced by coal plants that are over 30 years old. Rather than closing this loophole, the Bush administration has announced measures that will give those dirty old clunkers a new lease on life. Nor are poor countries blameless: many subsidise electricity heavily in the name of helping poor people, but rich farmers and urban elites then get to guzzle cheap (mostly coal-fired) power.

That points to a second prescription: the rich world could usefully help poorer countries to switch to cleaner energy. A forthcoming study by the International Energy Agency estimates that there are 1.6 billion people in the world who are unable to use modern energy. They often walk many miles to fetch wood, or collect cow dung, to use as fuel. As the poor world grows richer in coming decades, and builds thousands of power plants, many more such unfortunates will get electricity. That good news will come with a snag. Unless the rich world intervenes, many of these plants will burn coal in a dirty way. The resultant surge in carbon emissions will cast a grim shadow over the coming decades. Ending subsidies for exporters of fossil-fuel power plants might help. But stronger action is probably needed, meaning that the rich world must be ready to pay for the poor to switch to low-carbon energy. This should not be regarded as mere charity, but rather as a form of insurance against global warming.

The final and most crucial step is to start pricing energy properly. At the moment, the harm done to human health and the environment from burning fossil fuels is not reflected in the price of those fuels, especially coal, in most countries. There is no perfect way to do this, but one good idea is for governments to impose a tax based on carbon emissions. Such a tax could be introduced gradually, with the revenues raised returned as reductions in, say, labour taxes. That would make absolutely clear that the time has come to stop burning dirty fuels such as coal, using today's technologies.

The dawning of the age of hydrogen

None of these changes need kill off coal altogether. Rather, they would provide a much-needed boost to the development of low-carbon technologies. Naturally, renewables such as solar and wind will get a boost. But so too would "sequestration", an innovative way of using fossil fuels without releasing carbon into the air.

This matters for two reasons. For a start, there is so much cheap coal, distributed all over the world, that poor countries are bound to burn it. The second reason is that sequestration offers a fine stepping-stone to squeaky clean hydrogen energy. Once the energy trapped in coal is unleashed and its carbon sequestered, energy-laden hydrogen can be used directly in fuel cells. These nifty inventions can power a laptop, car or home without any harmful emissions at all.

It will take time to get to this hydrogen age, but there are promising harbingers. Within a few years, nearly every big car maker plans to have fuel-cell cars on the road. Power plants using this technology are already trickling on to the market. Most big oil companies have active hydrogen and carbon-sequestration efforts under way. Even some green groups opposed to all things fossil say they are willing to accept sequestration as a bridge to a renewables-based hydrogen future.

Best of all, this approach offers even defenders of coal a realistic long-term plan for tackling climate change. Since he rejected the UN's Kyoto treaty on climate change, George Bush has been portrayed as a stooge for the energy industry. This week, California's legislature forged ahead by passing restrictions on emissions of greenhouse gases; a Senate committee has acted similarly. Mr Bush, who has made surprisingly positive comments about carbon sequestration and fuel cells, could silence the critics by following suit. By cracking down on carbon and embracing hydrogen, he could even lead.

A great wall of waste

China is slowly starting to tackle its huge pollution problems
BEIJING, GUANGZHOU, HONG KONG, SHANGBA, SHANGHAI and TAIHE

PLUGGING a cigarette into his mouth, He Shouming runs a nicotine-stained fingernail down a list of registered deaths in Shangba, dubbed "cancer village" by the locals. The Communist Party official in this cluster of tiny hamlets of 3,300 people in northern Guangdong province, he concludes that almost half the 11 deaths among his neighbours this year, and 14 of the 31 last year, were due to cancer.

Mr He blames Dabaoshan, a nearby mineral mine owned by the Guangdong provincial government, and a host of smaller private mines for spewing toxic waste into the local rivers, raising lead levels to 44 times permitted rates. Walking around the village, the water in the streams is indeed an alarming rust-red. A rice farmer complains of itchy legs from the paddies, and his wife needs a new kettle each month because the water corrodes metal. "Put a duck in this water and it would die in two days," declares Mr He.

Poisons from the mines are also killing the village's economy, which depends on clean water to irrigate its crops, says Mr He. Rice yields are one-third of the national average and nobody wants to buy the crop. Annual incomes here have been stuck at less than 1,500 yuan ($180) per person for a decade, almost three times lower than the average in Guangdong province. The solution to Shangba's nightmare would be a local reservoir, but that idea was abandoned after various tiers of government squabbled over the 8.4m yuan cost.

Some 200km (124 miles) farther south and several decades into the future sits the Taihe landfill plant. Built for 540m yuan by Onyx, a waste-management company that is part of Veolia, a French utility, it has handled all of Guangzhou city's solid waste for the past two years. Each hour 140 trucks snake into the site, bringing 7,000 tonnes of rubbish a day from the 9.9m inhabitants of Guangdong's capital. In October delegates from 300 other municipalities will visit Taihe, promoted by central government as a role model of technology.

Smart cards record each truck's load, since Onyx charges by weight. Unrippable German fabric lines the crater into which the waste is dumped, stopping leachate—a toxic black liquid—from leaking into the groundwater, as it does at almost all Chinese-run sites. Most landfill in China is wet (solid rubbish, such as old TVs, is scavenged), and the Taihe plant collects a full 1,300 tonnes of the black liquid daily. Chemical and filtration systems to neutralise it are its biggest cost. Expensive too is the extraction equipment to gather another by-product, methane gas, which Onyx plans to feed into generators that will supply electricity to the local grid. Finally, the waste is topped off with plastic caps, deodorised and landscaped, while a crystal-clear fountain at the entrance tinkles with the cleaned-up leachate.

The extremes represented by Shangba and Taihe explain why it is difficult to get an accurate picture of China's pollution. In a country where data are untrustworthy, corruption rife and the business climate for foreigners unpredictable, neither the cause of Shangba's problems nor the smooth efficiency of Taihe are necessarily what they seem. As with many other aspects of China's economic development, rapid progress and bold experiments in some areas are balanced by bureaucratic rigidity and stagnation in others.

Certainly, awareness of China's environmental problems is rising among policymakers at the highest level—reflected in a new package of right-sounding initiatives like a "green GDP" indicator to account for environmental costs. So is the pressure, both internal and international, to fix them. But while all developing economies face this issue, there are historical, political and institutional reasons why it will be a long and complicated process in China. There is some cause for optimism, not least an influx of foreign technology and capital. But progress on pollution is unlikely to be as rapid or uniform as the government and environmentalists desire.

Nor should it necessarily be. China's need to lift so many people out of poverty (the country's average annual income per head has only just breached $1,000), holds the edge over long-term considerations like sustainable development. The priorities of environmental activists, both foreign and Chinese, almost never reflect this. Greenpeace lobbies for China to invest in wind farms, an unrealistic answer to the

41

country's power needs, while environmentalists from rich countries naively tell aspiring Chinese to eschew their new cars and air-conditioners.

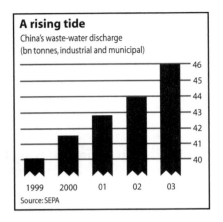

A rising tide
China's waste-water discharge
(bn tonnes, industrial and municipal)

Source: SEPA

Nor any drop to drink

That is not to deny the huge scale of China's environmental challenges. Water and waste pollution is the single most serious issue. Pan Yue, deputy head of the State Environmental Protection Administration (SEPA), the country's environmental watchdog ministry, calls it "the bottleneck constraining economic growth in China". Per head, China's water resources are among the lowest in the world and concentrated in the south, so that the north and west experience regular droughts. Inadequate investments in supply and treatment infrastructure means that even where water is not scarce, it is rarely clean. Around half the population, or 600m people, have water supplies that are contaminated by animal and human waste.

In late July an environmental disaster occurred on the Huai river, one of China's seven big rivers. A 133km-long black and brown plume swept along the river killing millions of fish and devastating wildlife. According to Mr Pan, the catastrophe occurred because too much water had been taken from the river system, reducing its ability to clean itself. Others say that numerous factories dump untreated waste directly into the water.

As for used water, with a national daily sewage rate of around 3.7 billion tonnes, China would need 10,000 waste-water treatment plants costing some $48 billion just to achieve a 50% treatment rate, according to Frost & Sullivan, a consultancy. SEPA found over 70% of the water in five of China's seven major river systems was unsuitable for human contact. As more people move into cities, the problem of household waste is becoming severe. Only 20% of China's 168m tonnes of solid waste per year is properly disposed of.

The air is not much better. "If I work in your Beijing, I would shorten my life at least five years," Zhu Rongji told city officials when he was prime minister in 1999. According to the World Bank, China has 16 of the world's 20 most polluted cities. Estimates suggest that 300,000 people a year die prematurely from respiratory diseases.

The main reason is that around 70% of China's mushrooming energy needs are supplied by coal-fired power stations, compared with 50% in America. Combined with the still widespread use of coal burners to heat homes, China has the world's highest emissions of sulphur dioxide and a quarter of the country endures acid rain. In 2002, SEPA found that the air quality in almost two-thirds of 300 cities it tested failed World Health Organisation standards—yet emissions from rocketing car ownership are only just becoming an issue. Hopes that China will "leapfrog" the West with super-green cars are naive, since dirty fuel messes up clean engines and the high cost of new cars keeps old ones on the road. Sun Jian, the second-ranking official at Shanghai's environmental protection bureau, estimates that 70% of Shanghai's 1m cars do not even reach the oldest European emission standards.

Farmland erosion and desertification resulted in Beijing being hit with 11 sandstorms in 2000, prompting Mr Zhu to wonder whether the advancing desert might force him to relocate the capital. A year later, the yellow dust clouds were so extensive

that they raised complaints in South Korea and Japan and travelled as far as America. A partial logging ban and massive replanting appear to have reversed China's deforestation, but its grass and agricultural land continue to shrink.

Adding it all up, the World Bank concludes that pollution is costing China an annual 8-12% of its $1.4 trillion GDP in direct damage, such as the impact on crops of acid rain, medical bills, lost work from illness, money spent on disaster relief following floods and the implied costs of resource depletion. With health costs escalating, that figure will increase, giving rise to some grim prognoses that growth itself will be undermined. "Ignored for decades, even centuries, China's environmental problems have the potential to bring the country to its knees economically," argues Elizabeth Economy, author of "The River Runs Black", a new book on China's pollution.

SEPA's Mr Pan is gloomier still: "Our natural resources will soon be unable to support our population." His predecessor Qu Geping, the first head of China's National Environmental Protection Agency (SEPA's forerunner) in 1985, believes that while the official goal of quadrupling 2002 GDP by 2020 can be "healthily achieved", if nothing is done about the environment, economic growth could grind to a halt.

But China's relationship with its environment has long been uneasy. For centuries, the country's rulers subjugated their surroundings rather than attempting to live in harmony with them. Mao declared that man must "conquer nature and thus attain freedom from nature". In the past two decades, the toll extracted by China's manufacturing-led development and the sheer scale of its 9%-a-year economic expansion has only increased.

From conquest to nurture

This has spurred the government into belated action. In 1998, Mr Zhu elevated SEPA to ministerial rank and three years later the 10th Five-year

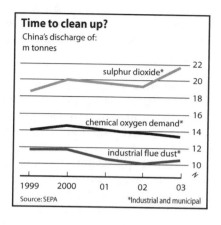

Time to clean up?
China's discharge of:
m tonnes

sulphur dioxide*

chemical oxygen demand*

industrial flue dust*

1999   2000   01   02   03
Source: SEPA   *Industrial and municipal

Plan for Environmental Protection set ambitious emission-reduction targets and boosted environmental spending to 700 billion yuan ($85 billion) for 2001–05—equivalent to 1.3% of GDP, up from 0.8% in the early 1990s (though still below the 2% suggested by the World Bank). A legal framework has been created. And the rhetoric has changed too, with Hu Jintao and Wen Jiabao, the current president and prime minister, now stressing balanced development rather than all-out economic growth.

Beijing's good intentions, however, have so far had only limited impact, thanks to the vast, decentralised bureaucracy through which it is forced to govern such a huge country. As Ken Lieberthal, a China expert at the University of Michigan, explains: "Much of the environmental energy generated at the national level dissipates as it diffuses through the multi-layered state structure, producing outcomes that have little concrete effect."

SEPA, the government's chosen weapon in the fight against pollution, is under-resourced despite its enhanced status, with little money and just 300 central staff. In the capital, it must battle for influence with other agencies, such as the Construction Ministry that handles water and sewage treatment. Bureaucratic rivalries mean there is no co-operation and no sharing of the (often patchy) data that are collected with limited funds, observes Bruce Murray, the Asian Development Bank's representative in China.

Around the country, SEPA's branches, known as Environmental Protection Bureaus, are supposed to monitor pollution, enforce standards and collect fines. But they are more in thrall to local governments—whose priorities are to maintain growth and employment in their jurisdiction—than to head office in Beijing. It is no rarity, therefore, to find a bureau imposing a fine on a dirty local enterprise (thus fulfilling its duty), but then passing the money on to the local administration, which refunds it to the company via a tax break. "The environmental management system needs real reform," says Ma Jun, an environmental scholar. "The bureaus depend on the local government for their salaries and pensions. How can they enforce regulations against the local government?" Mr Pan complains that SEPA cannot effectively push through central edicts because it does not directly employ environmental personnel at the local level. Mr Sun at the Shanghai bureau says that SEPA has given him only 300 people with which to police 20,000 factories.

Make polluters pay

SEPA's impotence is one reason why penalties, even when it can impose them, remain laughably light. Mr Sun says the maximum he can fine a polluting company in Shanghai—a model city when it comes to the environment—is 100,000 yuan or about $12,000. But just as fundamental is that China lacks an understanding of the concept that the polluter should pay. "The legacy of the old, centrally planned economy is that electricity and water are treated as free goods or goods to be provided at minimal cost," says the ADB's Mr Murray. Since the utilities cannot pass on the costs of cleaner water or lower power-station emissions to consumers, they fight any drive for higher standards and conservation tooth and nail. Even the central government is unwilling to impose price rises in basic services that could spark public unrest.

Water is an example. While customer tariffs have been raised in showcase cities, such as Beijing and Dalian in the north-east, water remains stunningly cheap in China. According to the World Bank, water for agriculture, which makes up three-quarters of the total used, is priced at 0.03 yuan (0.4 cents) per cubic metre, or about 40% of cost. More than half is lost in leaky irrigation systems. Meanwhile, the cost of more modern services, such as Guangzhou's solid-waste disposal, is entirely borne by the government.

Without the introduction of realistic pricing, China will not be able to afford to clean up its pollution, particularly the cost of enough foreign technology. Yet a system allocating the costs to the polluter will be hard to introduce and enforce. Even in Hong Kong, the territory's environment minister Sarah Liao concedes there is no tradition of having consumers bear the full costs of environmental regulations.

Ms Liao can also testify to the mainland's ambivalent attitude when it comes to letting outsiders help. She started looking into how the Pearl River delta's pollution was affecting Hong Kong back in 1999, but her requests to start monitoring emissions were repeatedly rebuffed even when she offered to pay for the equipment. Data collection finally started this year. For Thames Water, a British utility that is now a part of Germany's RWE, the experience was much worse. In June, Thames pulled out of a $73m advanced waste-water treatment plant it had built and was running in Shanghai, after the central government ruled that the fixed annual 15% return it had negotiated was now illegal.

There is no need to be unremittingly gloomy about China's environment, nevertheless. As developing countries get richer, they tend to pollute less. Nationally in China, discharges of chemical oxygen have declined over the past three years, those of industrial dust have stabilised and sulphur-dioxide emissions had been on the downtrend until

2003 when energy shortages increased demand for sulphurous coal (see charts). Most east-coast cities are enjoying more sunny days and the pollution load in the rivers is falling. Environmentally, in many places, China may have passed its nadir.

The government is increasing environmental spending and the more concerned attitude of the top leadership could filter down the hierarchy if the performance of officials starts being measured partly on environmental criteria, as Mr Qu hints it might. But the bigger incentive is that Beijing is under pressure to do more, partly from domestic public opinion. As urban Chinese see their material wealth increase, more are caring about the environment, while the concerns of the poor are increasingly being channelled by green non-governmental organisations. Though these remain extremely weak—few have more than a handful of members and all need government affiliation—Mr Wen said recently he suspended plans for the construction of 13 dams along the Nu river in Yunnan province partly because of the concerns outlined by such groups.

External pressure is even greater. Despite reservations, foreign companies are flocking to China, scenting a fast-growing market for their environmental technologies and skills. International agencies are tying funds to environmental criteria, while foreign governments are beginning to complain about China's dust storms and greenhouse-gas emissions. All this will help spread best practices. Beijing is fast cleaning up ahead of the 2008 Olympics, moving out factories and introducing clean-vehicle technology: a new premium is being placed on global respectability.

Of course, environmental problems and their huge costs will dog China for many years. In a country where the public is not free to speak, too many courts are toothless and environmental groups remain on a tight leash, it will be hard to know if the government's avowedly green policies are being implemented. But China deserves credit for its attempts to clean itself up. The balance between sustainable development and economic growth will have to be continuously adjusted in the future. Right now, China is probably moving in the right direction.

The new coal rush

Rising energy needs and relaxed regulations fuel a resurgence

FRANK JOSSI

Turn on a light in Vancouver, Canada, a few years from now and the power just may originate from a coal-fired power plant in tiny Hardin, Mont. The businesses and homes of the Twin Cities region may sometime trace the source of their power to a unique coal "gasification" plant in northern Minnesota that may require congressional support.

As demand for electricity increases and the price of natural gas skyrockets, utility companies in and near the Ninth District have begun to turn to coal as a source of power for the future. The proposals range from small plants to huge facilities hoping to cash in on growing metropolitan markets like the Twin Cities and those of the Pacific Northwest.

Coal has become a hot commodity, a potential answer to the nation's insatiable desire for inexpensive power and a potential economic boost for remote towns that welcome the industry with open arms. Coal offers a large reservoir of reserves and provides a more accessible and dependable power supply than gas, wind or solar energy, and a less dangerous one than nuclear power, supporters say.

Huge coal-fired plants are being proposed in small towns throughout North Dakota and Montana, as well as near more populous Wausau, Wis. A liberal count puts the number of publicly announced coal-fired plant proposals in or very near the district at six in Montana, two in North Dakota, two in northern and western Wisconsin, and one each in Minnesota and South Dakota. Even the Upper Peninsula of Michigan is getting in on the coal action, boasting an intriguing proposal that would use a federal government loan to install a new coal cleansing process that would be added to an existing plant

Hot opportunities

Environmentalists and the media report that utilities have made more than 100 proposals for new coal-fired plants across the nation since 2000 at a total cost of more than $72 billion.

"There's no question about it, there's been a resurgence in interest in coalfired power plants," said Jerry Va-

ninetti, president of Great Northern Power Development Co. of Denver, which has proposed two 500-megawatt coal-fired plants, one each in Montana and North Dakota. "It's the best reasonable option we have. The natural gas market has had great volatility and a tripling of prices in the last few years ... combine that with the fact there's been no building of nuclear plants recently and you begin to see how coal has become the best option available."

> **Coal has become a hot commodity, a potential answer to the nation's insatiable desire for inexpensive power and a potential economic boost for remote towns that welcome the industry with open arms.**

Though Minnesota has but one concrete proposal on the energy table, it likely represents only the beginning of a new age for coal in the state. Kandace Olsen, spokeswoman for Elk River, Minn.-based Great River Energy, said she has heard that nearly every major utility in the state—including Great River—is conducting studies to determine when and where its next coal-fired plant should be built "Everyone has something in the works," she said. "We've been talking to other utilities about projects but nothing's definite."

Assisting coal's resurgence as an energy source—after dropping off as utilities began building cleaner-burning natural gas plants—has been the Bush administration's relaxation of environmental standards and, ironically, several new technologies that allow for coal-fired power plants to burn cleaner and emit less mercury than in the past.

Relatively recent techniques such as coal gasification and "circulating fluidized bed" technology—a process employed in many European countries—promise to sequester and reduce pollutants. Minnesota's proposed coal gasification plant in Hoyt Lakes, at an abandoned taconite mine, has received unusually favorable regulatory treatment from the state, but its owners are still awaiting a controversial $800 million federal loan guarantee in the energy bill stalled in Congress at the time of this writing.

The cause has been aided by the U.S. Energy Department's Clean Coal Power Initiative, which offers interest-free loans to innovative projects. Coal also received a boost when the East Coast blackout last year called attention to the nation's geriatric transmission grid and aging fleet of power plants. Piecemeal deregulation at the state and federal level has likely induced a bit of speculative building by the industry as well.

Still, the industry faces heavy criticism and scrutiny from an environmental community concerned over carbon dioxide, sulfur dioxide and nitric oxide pollution, as well as mercury. More coal-fired plants—even clean-burning ones—may only aggravate global warming, they argue. Critics charge that the utility industry wants to "acquire federal permits for coal-fired power plants before it loses control of the White House to a more public-health-friendly administration," according to the Web site of GreenWatch, an environmental watchdog group.

Nor does a mere proposal make a plant. Some proposals remain highly speculative, often based on factors such as whether the country's transmission grid can be upgraded to efficiently deliver power generated in rural towns. Getting the appropriate permits, environmental and otherwise, usually takes years and often includes lawsuits, and construction of a coal-fired plant itself requires at least four years.

If the past is prologue, those 100 proposals nationwide will be whittled down dramatically due to competition from other energy sources, costly and time consuming lawsuits, changing federal priorities and other complex hurdles faced by the energy industry.

Why coal, why now?

Most coal-fired plants in the nation and the district are between 20 and 30 years old, according to state and utility officials. A construction boom in coal-fired plants in the late 1970s and early 1980s left Minnesota and much of the Upper Midwest "overbuilt" and with more than enough power, said Olsen. During that time Great River built the 1,200-megawatt Coal Creek Station 50 miles north of Bismarck, the largest power plant in North Dakota, that required a high-voltage, direct-current transmission line, which drew many protesters.

"Now we used up that capacity," Olsen said, adding that coal remains "the best fuel" for larger "baseload" energy plants that supply 24/7 energy to meet the daily needs of consumers.

Blessed with a mother lode of coal, Montana and North Dakota have become hotspots for generation activity. "Montana has the biggest coal reserves of any state in the country," said Vaninetti. "There's enough coal in that state to serve the country for 12,800 years at current production rates." Add North Dakota to the equation, he

said, and the two states together have enough coal to replace the nation's entire energy infrastructure—hydroelectric, nuclear, wind and solar—with enough fuel to last 2,800 years.

Unlike in the other states in the region, new coal-fired plants in these two states will export as much as two-thirds of their energy into transmission grids bound for the Twin Cities, the Pacific Northwest and Canada. Centennial Energy Resources of Hardin, Mont., will sell its energy to BC Hydro, one of Canada's largest utilities.

Montana and North Dakota have different approaches to energy production based largely on the quality of their coal. With coal mining industry employment ranked sixth in the country in 2002, Montana boasts several rich veins of sub-bituminous coal in the Powder River Basin it shares with Wyoming. (The region also boasts coalbed methane, covered in the January 2003 issue of the *fedgazette*, online at www.minneapolisfed.org.)

Although recent statistics show that Montana exports more than 70 percent of its coal production annually, the new plants would export power, not coal, using "mine-mouth" operations at several locations. Mine-mouth combinations, already popular in North Dakota, place the power plant next to the mine. North Dakota's brand of coal, called lignite, holds a third of its weight in water, making the transport of it uneconomical, said Jim Deutsch, director of the reclamation division at North Dakota's Public Service Commission. The state currently ranks 10th in coal production, he said, all of it used to power the state's seven electricity plants.

With coal mining industry employment ranked sixth in the country in 2002, Montana boasts several rich veins of sub-bituminous coal in the Powder River Basin it shares with Wyoming.

The potential of rising demand for coal-fired electricity has also fueled the ambitions of Sioux Falls, S.D.-based Dakota Minnesota & Eastern Railroad (DM&E), which wants to spend $2 billion to upgrade and build new lines connecting the coal-rich Powder River Basin of Wyoming with markets to the east. Bitterly opposed by many communities that fear the prospect of 100-car trains running through their towns daily, the railroad nonetheless has moved forward on two fronts this year by winning a court battle with South Dakota over the question of eminent domain and by securing a $233 million loan from the federal government.

Opponents have pointed out that Xcel Energy's decision to convert power plants in Minneapolis and St. Paul to natural gas will reduce the state's coal needs as much as 20 percent. Though DM&E's Chief Executive Officer Kevin Schaeffer has said the rail will move grain as well as coal, the economics will not work without coal. In a May 4 article, Schaeffer told the *Argus Leader* of Sioux

Falls that "everything requires the Powder River Basin project as an anchor."

Big industry, sort of

Coal mining and power generation is one of the largest employers in North Dakota, with more than 20,000 workers producing a $1.5 billion economic impact, according to statistics compiled by the St. Paul-based Partners for Affordable Energy, and represents around 15 percent of the state's economy. With coal such an instrumental supplier of jobs in both North Dakota and Montana, their governors, predictably, are doing all they can to promote growth.

North Dakota, for example, created a private-public partnership called Lignite Vision 21, which helps subsidize the study of power plant proposals by utilities. And Gov. John Hoeven has aggressively pursued a deal with the Environmental Protection Agency, nearly five years in the making, which changes the way pollution from coal-fired plants will be measured, to facilitate the building of a coal-fired plant in South Heart, near Theodore Roosevelt National Park and the Lostwood National Wildlife area. (The federal EPA requires air quality standards four times more stringent near national parks than in other areas.)

Montana Gov. Judy Martz's Web site says she reduced "regulatory burdens" on generators and passed a "tax holiday" on coal-fired plants during her term. She has argued that Montana has $4 billion in untapped coal wealth. The federal government also agreed to transfer to the state land in the Otter Creek area boasting 533 million tons of coal, enough potential to attract the attention of a consortium of companies studying the potential of a 3,000-megawatt plant.

For all the talk of economic development, coal-fired plants no longer employ many people. A $300 million plant in Gascoyne, N.D., proposed by Montana-Dakota Utilities Co. offers work to only 100 people; the Hardin plant (built by an MDU subsidiary) will employ 35 people.

"You don't need a lot of people to run them anymore," said Dan Sharp, public relations manager for the company. "It's more about maintenance and monitoring in the plants. No one's shoveling coal anymore."

In a small town, however, even a small employment gain produces a noticeable ripple effect throughout the economy. Vaninetti argued that the two coal-fired plants proposed by Great Northern might employ only 150 people at each location but will have a multiplier effect on local employment "There are a lot of jobs created in remote areas of the country which embrace power plant development," he said.

Rural towns remain hungry for growth. "Small towns are dying to have power plants built there," said Steve Wegman of the South Dakota Public Utilities Commission. "The best place to build a power plant is a small town in South Dakota or Wisconsin or Kansas or Ne-

braska. No one wants power plants in cities, but in rural towns they actually want the plants," Wegman said.

Marketplace and environmental challenges

There are two major stumbling blocks for more coal-generated power: transmission hurdles and environmental opposition.

The more tangible of the two problems is transmission. Regardless of how much power a plant generates, it all needs to be delivered somewhere else, and existing transmission capacity is badly strained. Currently, energy generated in the Dakotas, Minnesota and Wisconsin goes to the Eastern power grid, upon which energy in much of the Ninth District is transported. Montana splits the difference, with the eastern half sending power to the Eastern grid, the western portion to the Western grid.

Could power from North Dakota's new plants be transported to the East by today's transmission lines that carry power across the country? "No, definitely not," said Jeff Webb, director of planning for the Midwest Independent Transmission System Operator, a Carmel, Ind.-based nonprofit trade group promoting bulk power distribution. North Dakota's ability to transmit power is at capacity now—19,500 megawatts—and even planned upgrades will not increase that amount substantially in the next five years, he said.

> ### "The best place to build a power plant is a small town in South Dakota or Wisconsin or Kansas or Nebraska. No one wants power plants in cities, but in rural towns they actually want the plants." —Steve Wegman

Nor is electricity demand in potential export markets like the Twin Cities and Wisconsin not currently being met, said Webb. Since power plants take years to receive approvals and get built, he suggested that some proposals are banking on transmission capacity being added around the same time utilities finish building and begin moving power.

Energy companies are not sitting idle. Webb pointed out that utilities and transmission companies in the Dakotas and Minnesota have formed the Upper Great Plains Transmission Coalition to address the issue. MDU's Sharp said his company has embarked upon a study to look at how transmission can be increased.

But adding transmission comes with its own challenges. A good example is the Arrowhead transmission line proposed five years ago from Duluth, Minn., to Wausau, which just got under way at a cost of $420 million. It still faces legal issues and further negotiations for the purchase of rights of way in Wisconsin.

Charlie Severance, manager of supply and wholesale services for Wisconsin Public Service Corp. (WPS) in Green Bay, said building a baseload plant like one

Proposed Ninth District Coal-Fired Plants

MICHIGAN

Location: Marquette
Sponsoring Utilities: WE Energies
Status: Received a Clean Coal Initiative grant to develop a process to isolate mercury, nitrogen dioxide and sulfur dioxide in an existing plant
Cost: $25 million, paid by federal government

MINNESOTA

Location: Hoyt Lakes
Sponsoring Utilities: Mesaba Energy, part of Excelsior Energy
Status: Uncertain, proposed 531-megawatt plant awaiting loan decision in pending Energy Bill
Cost: $1 billion

MONTANA

Location: Circle/Miles City
Sponsoring Utilities: Great Northern Power Development
Status: Proposed 560-megawatt plant
Cost: $850 million to $900 million

Location: Great Falls
Sponsoring Utilities: Southern Montana Electric Generation and Transmission Co-op
Status: Proposed 250-megawatt plant
Cost: $470 million

Location: Hardin
Sponsoring Utilities: Centennial Energy Resources LLC (subsidiary of Montana-Dakota Utilities Co.)
Status: 116-megawatt project now under construction
Cost: Undetermined

Location: Otter Creek
Sponsoring Utilities: Kennecott, Bechtel, Wesco
Status: Proposed 3,000-megawatt plant
Cost: Undetermined

Location: Roundup
Sponsoring Utilities: Bull Mountain Power
Proposal: 760-megawatt coal-fired plant
Status: Under review after lawsuits by environmentalists
Cost: $910 million

NORTH DAKOTA

Location: Gascoyne
Sponsoring Utilities: Montana-Dakota Utilities Co. and West moreland Power Co.
Status: Proposed 175-megawatt coal-fired lignite plant
Cost: $300 million

Location: South Heart
Sponsoring Utilities: Great Northern Power Development
Status: Proposed 500-megawatt coal-fired plant/60 megawatts of wind
Cost: $850 million to $900 million

SOUTH DAKOTA

Location: Edgemont Sponsoring Utilities: Kfx, Denver, Colo.
Status: Newly proposed, employs process eliminating water and heavy metals
Cost: $250 million

WISCONSIN

Location: Wausau
Utility: WPS Resources Corp.
Proposal: 515-megawatt coal-fired plant
Cost: $750 million

Location: Undetermined
Utilities: Alliant Energy Corp., WPS Resources Corp.
Proposal: 500-megawatt coal-fired plant
Cost: Undetermined

planned by his company outside Wausau is much easier than adding transmission lines since they involve "several hundreds of miles which effectively have thousands of landowners who could see their property's value drop. There's a lot of friction in building transmission lines."

WPS ought to know. The company is one of the backers of the Duluth-Wausau line and plans to tie it into a new 515-megawatt, $750 million coal-fired plant on the Wisconsin River south of Wausau called "Weston 4." As the name implies, three other coal-fired plants exist on the site, though none as big as the planned fourth one. Even with a new plant and the power transmission line. Severance said be sees Wisconsin's power grid as "gummed up" and likely in need of additional capacity in the near future.

The inability to transmit power should not be underestimated. Over the past four years the majority of plants added to the nation's fleet of energy generators have been gas-fired, including several on the Gulf Coast that are not in production due to transmission challenges, said Ellen Vancko, director of communications for Princeton, N.J.-based North American Electric Reliability Council (NERC). " 'Build it and they will come' didn't work because in the Southeast on the Gulf Coast they can't get the energy out," she said. "It's a big issue."

Vancko does not know whether the plants of the Ninth District are economically feasible, but NERC's own research suggests "capacity margins"—the difference between generation and need—are "declining after 2005." That may indicate a need for new plants, she said. On the other hand, NERC's data show power supplies available now will be more than enough to handle summer and winter spikes this year.

Black mark for environment

Since coal-fired plants represent more than a third of the nation's CO_2 emissions—the rest comes from vehicles and other sources—coal's impact on the environment will have to be addressed at some future point by the nation's utilities, said Michael Noble, executive director of Minnesotans for an Energy-Efficient Economy. Though the Bush administration backed out of the CO_2-reducing Kyoto Treaty, it is only a matter of time before the United States will be forced to seriously propose a national policy to begin eliminating CO_2, he said.

Caps on CO_2 emission have wide support in the U.S. Senate and even among oil companies such as British Petroleum and Shell, said Noble. Investors in utilities without carbon emission policies could take a financial bath in the future when, inevitably, international pressure on the United States eventually leads to a national policy of capping coal-fired plant emissions.

Noble points out the new coal-fired plants do burn cleaner and meet "modern pollution standards." But the existing fleet of plants requires upgrades to remove CO_2 emissions, he added, and many are unlikely to be modernized anytime soon after the Bush Administration dropped allegations of Clean Air Act violations against dozens of coal-fired power plants late last year.

Coal-fired plants have been fingered as the primary source of mercury emission in many streams and lakes, said Bill Grant, associate executive director of the Midwest Izaak Walton League. Mercury represents a health hazard for child-bearing women since it can cause neurological disorders in infants, he said. While new technologies can remove a great deal of mercury, he conceded, whether the new plants will employ them remains a question, especially since the EPA late last year changed mercury standards. (The Sierra Club has sued the EPA over the matter.)

In Montana, a proposal for a new coal-fired power plant in Roundup, a city of about 2,000 located roughly in the middle of the state, has generated a backlash for being located so near Yellowstone National Park, according to Jeanne Charter, a rancher and member of the Northern Plains Resource Council.

But there are other pressures moving against the proposal as well. Existing coal plants already put stress on the state's livestock, Charter said. Pollution from coal-fired plants results in acid rain that hurts vegetation on farms and ranches, and those same power plants often require great amounts of water not easily available in the drought stricken West. "This kind of development competes with agricultural uses of water," she said.

Charter also questioned the ability to transmit power outside the region. "The biggest problem is there are no big power lines in that area," she said.

Ultimately environmental and transmission obstacles torpedo some projects, and with them, a portion of the coal market. Earlier this year, Dairyland Power Cooperative of La Crosse, Wis., had publicly announced the construction of a potential power plant in one of three small communities in Wisconsin and Iowa. The town of Alma, Wis., was more than willing to have a power plant constructed there and welcomed the utility's interest, according to news reports. Yet Dairyland quickly scotched that plan and now will likely invest in the construction of a 500-megawatt power plant in Wausau being proposed by WPS, said Deb Mirasola, communications director for Dairyland. "We're evaluating our options," she said. "We're still in negotiations, and we'll likely announce more decisions toward the end of summer and early fall."

Great River Energy, recalled Olsen, used money from the Lignite Vision 21 project three years ago to look at adding a $700 million plant in North Dakota. A year later the company dropped the plan and refunded the state $500,000 for the cost of the study. It decided to build a "peaking" plant, a smaller natural gas plant that contributes energy in high demand seasons like summer and winter.

In a press release Great River's managers cited the issue of transmission as a deal killer. Moving forward, other coal-fired plants might find the same challenge too daunting to overcome.

UNIT 3
The Region

Unit Selections

Key Points to Consider

- To what regions do you belong?

- Why are maps and atlases so important in discussing and studying regions?

- What major region in the world is experiencing change? Which ones seem not to change at all? What are some reasons for the differences?

- What are the long-term impacts of drought in the western U.S.?

- Why are regions in Africa suffering so greatly?

- Will agricultural output keep pace with population growth in the 21st century?

- Why is regional study important?

Student Website
www.mhcls.com/online

Internet References
Further information regarding these websites may be found in this book's preface or online.

AS at UVA Yellow Pages: Regional Studies
  http://xroads.virginia.edu/~YP/regional.html
Can Cities Save the Future?
  http://www.huduser.org/publications/econdev/habitat/prep2.html
NewsPage
  http://www.individual.com
World Regions & Nation States
  http://www.worldcapitalforum.com/worregstat.html

The region is one of the most important concepts in geography. The term has special significance for the geographer, and it has been used as an area classification system in the discipline.

Two of the regional types most used in geography are "uniform" and "nodal." A uniform region is one in which a distinct set of features is present. The distinctiveness of the combination of features marks the region as being different from others. These features include climate type, soil type, prominent languages, resource deposits, and virtually any other identifiable phenomenon having a spatial dimension.

The nodal region reflects the zone of influence of a city or other nodal place. Imagine a rural town in which a farm-implement service center is located. Now imagine lines drawn on a map linking this service center with every farm within the area that uses it. Finally, imagine a single line enclosing the entire area in which the individual farms are located. The enclosed area is defined as a nodal region. The nodal region implies interaction. Regions of this type are defined on the basis of banking linkages, newspaper circulation, and telephone traffic, among other things.

This unit presents examples of a number of regional themes. These selections can provide only a hint of the scope and diversity of the region in geography. There is no limit to the number of regions; there are as many as the researcher sets out to define.

The first article highlights "brainpower" in India and the country's rapid economic growth. "Between the Mountains" documents the struggle between India and Pakistan over Kashmir. Coverage of the changing geopolitical balance in East Asia follows. Two related articles on the geopolitical importance of the Black Sea region are next. The growth of Central Washington's Hispanic population is discussed in this *Focus on Geography* article. The next article deals with the severe shortage of fresh water in China. "Living with the Desert" emphasizes the degree of care that must be taken to protect the sensitive and fragile desert landscape. The next selection argues convincingly for the care of the oceans and their international governance.

THE RISE OF **INDIA**

Growth is only just starting, but the country's brainpower is already reshaping Corporate America

By Manjeet Kripalani and Pete Engardio

Pulling into General Electric's John F. Welch Technology Center, a uniformed guard waves you through an iron gate. Once inside, you leave the dusty, traffic-clogged streets of Bangalore and enter a leafy campus of low buildings that gleam in the sun. Bright hallways lined with plants and abstract art—"it encourages creativity," explains a manager—lead through laboratories where physicists, chemists, metallurgists, and computer engineers huddle over gurgling beakers, electron microscopes, and spectrophotometers. Except for the female engineers wearing saris and the soothing Hindi pop music wafting through the open-air dining pavilion, this could be GE's giant research-and-development facility in the upstate New York town of Niskayuna.

It's more like Niskayuna than you might think. The center's 1,800 engineers—a quarter of them have PhDs—are engaged in fundamental research for most of GE's 13 divisions. In one lab, they tweak the aerodynamic designs of turbine-engine blades. In another, they're scrutinizing the molecular structure of materials to be used in DVDs for short-term use in which the movie is automatically erased after a few days. In another, technicians have rigged up a working model of a GE plastics plant in Spain and devised a way to boost output there by 20%. Patents? Engineers here have filed for 95 in the U.S. since the center opened in 2000.

Pretty impressive for a place that just four years ago was a fallow plot of land. Even more impressive, the Bangalore operation has become vital to the future of one of America's biggest, most profitable companies. "The game here really isn't about saving costs but to speed innovation and generate growth for the company," explains Bolivian-born Managing Director Guillermo Wille, one of the center's few non-Indians.

The Welch center is at the vanguard of one of the biggest mind-melds in history. Plenty of Americans know of India's inexpensive software writers and have figured out that the nice clerk who booked their air ticket is in Delhi. But these are just superficial signs of India's capabilities. Quietly but with breathtaking speed, India and its millions of world-class engineering, business, and medical graduates are becoming enmeshed in America's New Economy in ways most of us barely imagine. "India has always had brilliant, educated people," says tech-trend forecaster Paul Saffo of the Institute for the Future in Menlo Park, Calif. "Now Indians are taking the lead in colonizing cyberspace."

"Just like China drove down costs in manufacturing and Wal-Mart in retail, India will drive down costs in services," says an Indian IT exec

This techno take-off is wonderful for India—but terrifying for many Americans. In fact, India's emergence is fast turning into the latest Rorschach test on globalization. Many see India's digital workers as bearers of new prosperity to a deserving nation and vital partners of Corporate America. Others see them as shock troops in the final assault on good-paying jobs. Howard Rubin, executive vice-president of Meta Group Inc., a Stamford (Conn.) information-technology consultant, notes that big U.S. companies are shedding 500 to 2,000 IT staffers at a time. "These people won't get reabsorbed into the workforce until they get the right skills," he says. Even Indian execs see the problem. "What happened in manufacturing is happening in

WHERE INDIA IS MAKING AN IMPACT

SOFTWARE
India is now a major base for developing new applications for finance, digital appliances, and industrial plants.

IT CONSULTING
Companies such as Wipro, Infosys, and Tata are managing U.S. IT networks and re-engineering business processes.

CALL CENTERS
Thousands of Indians handle customer service and process insurance claims, loans, bookings, and credit-card bills.

CHIP DESIGN
Intel, Texas Instruments, and many U.S. startups use India as an R&D hub for mircroprocessors and multimedia chips.

...AND WHERE IT'S GOING NEXT

FINANCIAL ANALYSIS
Research for Wall street will surge as U.S. investment banks, brokerages, and accounting firms open big offices.

INDUSTRIAL ENGINEERING
India does vital R&D for GE Medical, GM, engine maker Cummins, Ford, and other manufactures plan big engineering hubs.

ANALYTICS
U.S. companies are hiring Indian math experts to devise models for risk analysis, consumer behavior, and industrial processes.

DRUG RESEARCH
As U.S. R&D costs soar, India is expected to be a center for biotechnology and clinical testing.

Data: *BusinessWeek*

services," says Azim H. Premji, chairman of IT supplier Wipro Ltd. "That raises a lot of social issues for the U.S."

No wonder India is at the center of a brewing storm in America, where politicians are starting to view offshore outsourcing as the root of the jobless recovery in tech and services. An outcry in Indiana recently prompted the state to cancel a $15 million IT contract with India's Tata Consulting. The telecom workers' union is up in arms, and Congress is probing whether the security of financial and medical records is at risk. As hiring explodes in India, the jobless rate among U.S. software engineers has more than doubled, to 4.6%, in three years. The rate is 6.7% for electrical engineers and 7.7% for network administrators. In all, the Bureau of Labor Statistics reports that 234,000 IT professionals are unemployed.

The biggest cause of job losses, of course, has been the U.S. economic downturn. Still, there's little denying that the offshore shift is a factor. By some estimates, there are more IT engineers in Bangalore (150,000) than in Silicon Valley (120,000). Meta figures at least one-third of new IT development work for big U.S. companies is done overseas, with India the biggest site. And India could start grabbing jobs from other sectors. A.T. Kearney Inc. predicts that 500,000 financial-services jobs will go offshore by 2008. Indiana notwithstanding, U.S. governments are increasingly using India to manage everything from accounting to their food-stamp programs. Even the U.S. Postal Service is taking work there. Auto engineering and drug research could be next.

More Science in Schools

TECH LUMINARY Andrew S. Grove, CEO of Intel Corp., warns that "it's a very valid question" to ask whether America could eventually lose its overwhelming dominance in IT, just as it did in electronics manufacturing. Plunging global telecom costs, lower engineering wages abroad, and new interactive-design software are driving revolutionary change, Grove said at a software confer-

WHY CORPORATE AMERICA IS BEATING A PATH TO INDIA

Growth prospects are accelerating...
GDP IN TRILLIONS OF U.S. DOLLARS
Data: Goldman, Sachs & Co.

...fed by booming exports of IT-related services...
BILLIONS OF U.S. DOLLARS
Annual Exports of Tech and Back-office Services
Data: McKinsey & Co. Nasscom

...a growing pool of tech graduates...
THOUSANDS
Engineers with College Degrees
Data: Nasscom estimates

...and a swelling workforce
MILLIONS
Indians Aged 15-54
Data: CLSA Emerging Markets, Statistical Outline of India

ence in October. "From a technical and productivity standpoint, the engineer sitting 6,000 miles away might as well be in the next cubicle and on the local area network." To maintain America's edge, he said, Washington and U.S. industry must double software productivity through more R&D investment and science education.

But there's also a far more positive view—that harnessing Indian brainpower will greatly boost American tech and services leadership by filling a big projected shortfall in skilled labor as baby boomers retire. That's especially possible with smarter U.S. policy. Companies from GE

WHO'S **BULKING** UP

Some of the biggest U.S. players in India

COMPANY	PURPOSE	INDIA STAFF
GE Captial Services	Back-office work	16,000
GE's John Welsh Tech Center	Product R&D	1,800
IBM Global Services	IT services, software	10,000
Oracle	Software, services	6,000**
EDS	IT services	3,500†
Texas Instruments	Chip design	900
Intel	Chip design, software	1,700
J.P. Morgan Chase	Back-office, analysis	1,200

*By 2005 **Unspecified †By 2004 Data: Company reports, Nasscom, Evalueserve

Medical Systems to Cummins to Microsoft to enterprise-software firm PeopleSoft that are hiring in India say they aren't laying off any U.S. engineers. Instead, by augmenting their U.S. R&D teams with the 260,000 engineers pumped out by Indian schools each year, they can afford to throw many more brains at a task and speed up product launches, develop more prototypes, and upgrade quality. A top electrical or chemical engineering grad from Indian Institutes of Technology (IITs) earns about $10,000 a year—roughly one-eighth of U.S. starting pay. Says Rajat Gupta, an IIT-Delhi grad and senior partner at consulting firm McKinsey & Co.: "Offshoring work will spur innovation, job creation, and dramatic increases in productivity that will be passed on to the consumer."

Whether you regard the trend as disruptive or beneficial, one thing is clear. Corporate America no longer feels it can afford to ignore India. "There's just no place left to squeeze" costs in the U.S., says Chris Disher, a Booz Allen Hamilton Inc. outsourcing specialist. "That's why every CEO is looking at India, and every board is asking about it." neoIT, a consultant advising U.S. clients on how to set up shop in India, says it has been deluged by big companies that have been slow to move offshore. "It is getting to a state where companies are literally desperate," says Bangalore-based neoIT managing partner Avinash Vashistha.

As a result of this shift, few aspects of U.S. business remain untouched. The hidden hands of skilled Indians are present in the interactive Web sites of companies such as Lehman Brothers and Boeing, display ads in your Yellow Pages, and the electronic circuitry powering your Apple Computer iPod. While Wall Street sleeps, Indian analysts digest the latest financial disclosures of U.S. companies and file reports in time for the next trading day. Indian staff troll the private medical and financial records of U.S. consumers to help determine if they are good risks for insurance policies, mortgages, or credit cards from American Express Co. and J.P. Morgan Chase & Co.

By 2008, forecasts McKinsey, IT services and back-office work in India will swell fivefold, to a $57 billion annual export industry employing 4 million people and accounting for 7% of India's gross domestic product. That growth is inspiring more of the best and brightest to stay home rather than migrate. "We work in world-class companies, we're growing, and it's exciting," says Anandraj Sengupta, 24, an IIT grad and young star at GE's Welch Centre, where he has filed for two patents. "The opportunities exist here in India."

If India can turn into a fast-growth economy, it will be the first developing nation that used its brainpower, not natural resources or the raw muscle of factory labor, as the catalyst. And this huge country desperately needs China-style growth. For all its R&D labs, India remains visibly Third World. IT service exports employ less than 1% of the workforce. Per-capita income is just $460, and 300 million Indians subsist on $1 a day or less. Lethargic courts can take 20 years to resolve contract disputes. And what pass for highways in Bombay are choked, crumbling roads lined with slums, garbage heaps, and homeless migrants sleeping on bare pavement. More than a third of India's 1 billion citizens are illiterate, and just 60% of homes have electricity. Most bureaucracies are bloated, corrupt, and dysfunctional. The government's 10% budget deficit is alarming. Tensions between Hindus and Muslims always seem poised to explode, and the risk of war with nuclear-armed Pakistan is ever-present.

So it's little wonder that, compared to China with its modern infrastructure and disciplined workforce, India is far behind in exports and as a magnet for foreign investment. While China began reforming in 1979, India only started to emerge from self-imposed economic isolation after a harrowing financial crisis in 1991. China has seen annual growth often exceeding 10%, far better than India's decade-long average of 6%.

In the Valley's Marrow

STILL, THIS DEEP SOURCE of low-cost, high-IQ, English-speaking brainpower may soon have a more far-reaching impact on the U.S. than China. Manufacturing—China's strength—accounts for just 14% of U.S. output and 11% of jobs. India's forte is services—which make up 60% of the U.S. economy and employ two-thirds of its workers. And Indian knowledge workers are making their way up the New Economy food chain, mastering tasks requiring analysis, marketing acumen, and creativity.

This means India is penetrating America's economic core. The 900 engineers at Texas Instruments Inc.'s Bangalore chip-design operation boast 225 patents. Intel Inc.'s Bangalore campus is leading worldwide research for the company's 32-bit microprocessors for servers and wireless chips. "These are corporate crown jewels," says Intel India President Ketan Sampat. India is even getting hard-wired into Silicon Valley. Venture capitalists say anywhere from one-third to three-quarters of the software, chip, and e-commerce startups they now back have Indian R&D teams from the get-go. "We can barely imagine investing in a company without at least asking what their plans are for India," says Sequoia Capital partner Michael Moritz, who nurtured Google, Flextronics, and Agile Software. "India has seeped into the marrow of the Valley [see box]."

It's seeping into the marrow of Main Street. This year, the tax returns of some 20,000 Americans were prepared by $500-a-month CPAs such as Sandhya Iyer, 24, in the Bombay office of Bangalore's MphasiS. After reading scanned seed and fertilizer invoices, soybean sales receipts, W2 forms, and investment records from a farmer in Kansas, Iyer fills in the farmer's 82-page return. "He needs to amortize these," she types next to an entry for new machinery and a barn. A U.S. CPA reviews and signs the finished return. Next year, up to 200,000 U.S. returns will be done in India, says CCH Inc. in Riverwoods, Ill., a supplier of accounting software. And it's not only Big Four firms that are outsourcing. "We are seeing lots of firms with 30 to 200 CPAs—even single practitioners," says CCH Sales Vice-President Mike Sabbatis.

A top electrical-engineering grad from one of the six Indian Institutes of Technology fetches about $10,000 a year

The gains in efficiency could be tremendous. Indeed, India is accelerating a sweeping reengineering of Corporate America. Companies are shifting bill payment, human resources, and other functions to new, paperless centers in India. To be sure, many corporations have run into myriad headaches, ranging from poor communications to inconsistent quality. Dell Inc. recently said it is moving computer support for corporate clients back to the U.S. Still, a raft of studies by Deloitte Research, Gartner, Booz Allen, and other consultants find that companies shifting work to India have cut costs by 40% to 60%. Companies can offer customer support and use pricey computer gear 24/7. U.S. banks can process mortgage applications in three hours rather than three days. Predicts Nandan M. Nilekani, managing director of Bangalore-based Infosys Technologies Ltd.: "Just like China drove down costs in manufacturing and Wal-Mart in retail," he says, "India will drive down costs in services."

GE Capital saves up to $340 million a year by performing some 700 tasks in India

But deflation will also mean plenty of short-term pain for U.S. companies and workers who never imagined they'd face foreign rivals. Consider America's $240 billion IT-services industry. Indian players led by Infosys, Tata, and Wipro got their big breaks during the Y2K scare, when U.S. outfits needed all the software help they could get. Indians still have less than 3% of the market. But by undercutting giants such as Accenture, IBM, and Electronic Data Systems by a third or more for software and consulting, they've altered the industry's pricing. "The Indian labor card is unbeatable," says Chief Technology Officer John Parkinson of consultant Cap Gemini Ernst & Young. "We don't know how to use technology to make up the difference."

Wrenching Change

MANY U.S. WHITE-COLLAR workers are also in for wrenching change. A study by McKinsey Global Institute, which believes offshore outsourcing is good, also notes that only 36% of Americans displaced in the previous two decades found jobs at the same or higher pay. The incomes of a quarter of them dropped 30% or more. Given the higher demands of employers, who want technicians adept at innovation and management, it could take years before today's IT workers land solidly on their feet.

India's IT workers, in contrast, sense an enormous opportunity. The country has long possessed some basics of a strong market-driven economy: private corporations, democratic government, Western accounting standards, an active stock market, widespread English use, and schools strong in computer science and math. But its bureaucracy suffocated industry with onerous controls and taxes, and the best scientific and business minds went to the U.S., where the 1.8 million Indian expatriates rank among the most successful immigrant groups.

Now, many talented Indians feel a sense of optimism India hasn't experienced in decades. "IT is driving India's boom, and we in the younger generation can really deliver the country from poverty," says Rhythm Tyagi, 22, a master's degree student at the new Indian Institute of Information Technology in Bangalore. The campus is completely wired for Wi-Fi and boasts classrooms with videoconferencing to beam sessions to 300 other colleges.

That confidence is finally spurring the government to tackle many of the problems that have plagued India for so long. Since 2001, Delhi has been furiously building a network of high-ways. Modern airports are next. Deregulation of the power sector should lead to new capacity. Free education for girls to age 14 is a national priority. "One by one, the government is solving the bottlenecks," says Deepak Parekh, a financier who heads the quasi-governmental Infrastructure Development Finance Co.

Future Vision

INDIA ALSO IS WORKING to assure that it will be able to meet future demand for knowledge workers at home and abroad. India produces 3.1 million college graduates a year, but that's expected to double by 2010. The number of engineering colleges is slated to grow 50%, to nearly 1,600, in four years. Of course, not all are good enough to produce the world-class grads of elite schools like the IITs, which accepted just 3,500 of 178,000 applicants last year. So there's a growing movement to boost faculty salaries and reach more students nationwide through broadcasts. India's rich diaspora population is chipping in, too. Prominent Indian Americans helped found the new Indian School of Business, a tie-up with Wharton School and Northwestern University's Kellogg Graduate School of Management that lured most of its faculty from the U.S. Meanwhile, the six IIT campuses are tapping alumni for donations and research links with Stanford, Purdue, and other top science universities. "Our mission is to become one of the leading science institutions in the world," says director Ashok Mishra of IIT-Bombay, which has raised $16 million from alumni in the past five years.

If India manages growth well, its huge population could prove an asset. By 2020, 47% of Indians will be between 15 and 59, compared with 35% now. The working-age populations of the U.S. and China are projected to shrink. So India is destined to have the world's largest population of workers and consumers. That's a big reason why Goldman, Sachs & Co. thinks India will be able to sustain 7.5% annual growth after 2005.

Skeptics fear U.S. companies are going too far, too fast in linking up with this giant. But having watched the success of the likes of GE Capital International Services, many execs feel they have no choice. Inside GECIS' Bangalore center—one of four in India—Gauri Puri, a 28-year-old dentist, is studying an insurance claim for a root-canal operation to see if it's covered in a certain U.S. patient's dental plan. Two floors above, members of a 550-strong analytics team are immersed in spreadsheets filled with a boggling array of data as they devise statistical models to help GE sales staff understand the needs, strengths, and weaknesses of customers and rivals. Other staff prepare data for GE annual reports, write enterprise resource-planning software, and process $35 billion worth of global invoices. Says GE Capital India President Pramod Bhasin: "We are mission-critical to GE." The 700 business processes done in India save the company $340 million a year, he says.

Indian finance whizzes are a godsend to Wall Street, too, where brokerages are under pressure to produce more independent research. Many are turning to outfits such as OfficeTiger in the southern city of Madras. The company employs 1,200 people who write research reports and do financial analysis for eight Wall Street firms. Morgan Stanley, J.P. Morgan, Goldman Sachs, and other big investment banks are hiring their own armies of analysts and back-office staff. Many are piling into Mindspace, a sparkling new 140-acre city-within-a-city abutting Bombay's urban squalor. Some 3 million square feet are already leased to Western finance firms. By yearend, Morgan Stanley will fill several floors of a new building.

For Silicon Valley startups, Indian engineers let them stretch R&D budgets. PortalPlayer Inc., a Santa Clara (Calif.) maker of multimedia chips and embedded software for portable devices such as music players, has hired 100 engineers in India and the U.S. who update each other daily at 9 a.m. and 10 p.m. J.A. Chowdary, CEO of PortalPlayer's Hyderabad subsidiary Pinexe, says the company has shaved up to six months off the development cycle—and cut R&D costs by 40%. Impressed, venture capitalists have pumped $82 million into PortalPlayer.

More Bang for the Buck

OLD ECONOMY COMPANIES are benefiting, too. Engine maker Cummins plans to use its new R&D center in Pune to develop the sophisticated computer models needed to design upgrades and prototypes electronically. Says International Vice-President Steven M. Chapman: "We'll be able to introduce five or six new engines a year instead of two" on the same $250 million R&D budget—without a single U.S. layoff.

The nagging fear in the U.S., though, is that such assurances will ring hollow over time. In other industries, the shift of low-cost production work to East Asia was followed by engineering. Now, South Korea and Taiwan are global leaders in notebook PCs, wireless phones, memory chips, and digital displays. As companies rely more on IT engineers in India and elsewhere, the argument goes, the U.S. could cede control of other core technologies. "If we continue to offshore high-skilled professional jobs, the U.S. risks surrendering its leading role in innovation," warns John W. Steadman, incoming U.S. president of Institute of

WHERE CHINA IS WAY AHEAD...	...WHERE INDIA HAS THE EDGE
GROWTH GDP has risen an average of 8% for the past decade, compared with India's 6%.	**LANGUAGE** English gives India a big edge in IT services and back-office work.
INFRASTRUCTURE Highways, ports, power sector, and industrial parks are far superior.	**CAPITAL MARKETS** Private firms have readier access to funding. China favors state sector.
FOREIGN INVESTMENT China lures $50 billion-plus a year. India gets $4 billion.	**LEGAL SYSTEM** Contract law and copyright protection are more developed than in China.
EXPORTS $266 billion reported in 2002 was more than four times India's total.	**DEMOGRAPHICS** Some 53% of India's population is under age 25, vs. 45% in China.

BRAINPOWER
India and Silicon Valley: Now the R&D Flows Both Ways

The ravages of the dot-com bust are still evident at Andale Inc.'s Mountain View (Calif.) headquarters. Half the office space sits abandoned, one corner of it heaped with discarded cubicle dividers and file cabinets. But looks are deceptive. The four-year-old startup, which offers software and research tools for online auction buyers and sellers, has seen its workforce nearly quadruple in the past year—with most of those jobs in Bangalore.

Andale's 155 workers in India, where employing a top software programmer runs a small fraction of the cost in the U.S., have been the key to the company's survival, says Chief Executive Munjal Shah, who grew up in Silicon Valley. In fact, Indian talent is adding vitality throughout Silicon Valley, where it's getting hard to find an info-tech startup that doesn't have some research and development in such places as Bangalore, Bombay, or Hyderabad. Says Shah: "The next trillion dollars of wealth will come from companies that straddle the U.S. and India."

The chief architects of this rising business model are the 30,000-odd Indian IT professionals who live and work in the Valley. Indian engineers have become fixtures in the labs of America's top chip and software companies. Indian émigrés have also excelled as managers, entrepreneurs, and venture capitalists. As of 2000, Indians were among the founders or top execs of at least 972 companies, says AnnaLee Saxenian, who studies immigrant business networks at the University of California at Berkeley.

Until recently, that brainpower mostly went in one direction, benefiting the Valley more than India. Now, this ambitious diaspora is generating a flurry of chip, software, and e-commerce startups in both nations, mobilizing billions in venture capital. The economics are so compelling that some venture capitalists demand Indian R&D be included in business plans from Day One. Says Robin Vasan, a partner at Mayfield in Menlo Park: "This is the way they need to do business."

The phenomenon is due in no small part to the professional and social networks Indians have set up in the Valley, such as The Indus Entrepreneurs (TiE), in Santa Clara: It now has 42 chapters in nine countries. Prominent Indians such as TiE founder and serial entrepreneur Kanwal Rekhi, venture capitalist Vinod Khosla, entrepreneur Kanwal Rekhi, and former Intel corp. executive Vin Dham serve as startup mentors and angel investors. In early November, Bombay-born Ash Lilani, senior vice-president at Silicon Valley Bank, led 20 Valley VCs on their first trip to India to scout opportunities. Of the bank's 5,000 Valley clients, 10% have some development work in India, but that's expected to rise to 25% in two years.

Such opportunities for the Valley's Indians flow both ways. Hundreds have returned to India since 2000 to start businesses or help expand R&D labs for the likes of Oracle, Cisco Systems, and Intel.

The downturn—and Washington's decision to issue fewer temporary work visas—accelerated the trend. At a Nov. 6 tech job fair in Santa Clara, hundreds of engineers lined up, résumés in hand, for Indian openings offered by companies from Microsoft Corp. to Juniper Networks Inc. "The real development and design jobs are in India," says Indian-born job-seeker Jay Venkat, 24, a University of Alabama electrical engineering grad.

The deeper, more symbiotic relationship developing between the Valley and India goes far beyond the "body shopping" of the 1990s, when U.S. companies mainly wanted low-wage software-code writers. Now the brain drain from India is turning into what Saxenian calls "brain circulation," nourishing the tech scenes in both nations.

Some Valley companies even credit India with saving them from oblivion. Web-hosting software outfit Ensim Corp. in Sunnyvale relied on its 100-engineer team in Bangalore to keep designing lower-cost new products right through the downturn. "This company would not survive a day if not for the operation in India," says CEO Kanwal Rekhi. Before long, India may prove as crucial to the Valley's success as silicon itself.

—By Robert D. Hof in Santa Clara, Calif., with Manjeet Kripalani in Bombay

Electrical & Electronics Engineers Inc. That could also happen if many foreigners—who account for 60% of U.S. science grads and who have been key to U.S. tech success—no longer go to America to launch their best ideas.

Information-technology services could soon account for 7% of India's GDP

Throughout U.S. history, workers have been pushed off farms, textile mills, and steel plants. In the end, the workforce has managed to move up to better-paying, higher-quality jobs. That could well happen again. There will still be a crying need for U.S. engineers, for example. But what's called for are engineers who can work closely with customers, manage research teams, and creatively improve business processes. Displaced technicians who lack such skills will need retraining; those entering school will need broader educations.

Adapting to the India effect will be traumatic, but there's no sign Corporate America is turning back. Yet the India challenge also presents an enormous opportunity for the U.S. If America can handle the transition right, the end result could be a brain gain that accelerates productivity and innovation. India and the U.S., nations that barely interacted 15 years ago, could turn out to be the ideal economic partners for the new century.

—With Steve Hamm in New York

LETTER FROM KASHMIR

BETWEEN THE MOUNTAINS

India and Pakistan are caught in a dangerous struggle over Kashmir.
But what do its people want?

BY ISABEL HILTON

When the French doctor François Bernier entered the Kashmir Valley for the first time, in 1665, he was astounded by what he found. "In truth," he wrote, it "surpasses in beauty all that my warm imagination had anticipated. It is not indeed without reason that the Moghuls call Kachemire the terrestrial paradise of the Indies." The valley, which is some ninety miles long and twenty miles across, is sumptuously fertile. Along its floor, there are walnut and almond trees, orchards of apricots and apples, vineyards, rice paddies, hemp and saffron fields. There are woods on the lower slopes of the surrounding mountains—sycamore, oak, pine, and cedar. The southern side is bounded by the Pir Panjal, not the highest mountain range in Asia but one of the most striking, rising abruptly from the valley floor. The northern boundary is formed by the Great Himalayas. At the heart of the valley lie Dal Lake and the graceful capital, Srinagar.

For Europeans, Kashmir became a locus of romantic dreams, inspiring writers like the Irish poet Thomas Moore, who didn't even need to visit it to understand its charms. "Who has not heard of the Vale of Cashmere," he wrote in 1817, "with its roses the brightest that earth ever gave." So seductive was this landlocked valley that, like a beautiful woman surrounded by jealous lovers, Kashmir attracted a succession of invaders, each eager to possess her.

Srinagar is a city of waterways, floating gardens, and lotus beds, spanned by nine graceful bridges.

The Moghuls established their control in the sixteenth century. Kashmir became the northern limit of their Indian empire as well as their pleasure ground, a place to wait out

the summer heat of the plains. They built gardens in Srinagar, along the shores of Dal Lake, with cool and elegantly proportioned terraces—with fountains and roses and jasmine and rows of chinar trees. The Moghul rulers were followed by the Afghans and, later, by the Sikhs from the Punjab, who were driven out in the nineteenth century by the British, who then sold the valley, to the abiding shame of its residents, for seven and a half million rupees to the maharaja, Gulab Singh. Singh was the notoriously brutal Hindu ruler of Jammu, the region that lay to the south, beyond the Pir Panjal, on the edge of the plains of the Punjab.

Under Singh, the Kashmir Valley was conjoined in the princely state of Jammu and Kashmir. According to one calculation of the purchase, the ruler of the newly formed state had bought the people of Kashmir for approximately three rupees each, a sum he was to recover many times over through taxation. For the maharaja and his descendants and their visitors, the valley was luxurious paradise; they enjoyed fishing and duck shooting, boating excursions on Dal Lake, picnics in the hills and the saffron fields, moonlit parties in the magnificent gardens. In the penetrating cold of the winters, the visitors, and the maharaja, left the valley to itself and returned to Jammu.

Kashmir was also a natural crossroads. The Silk Route, with its great camel trains from China, passed to the north, and the country's mountain passes opened routes to the Punjab, Afghanistan, and Jammu. Through them successive intruders brought different cultures that added layers to Kashmir's own. The Kashmiri language was a mixture of Persian, Sanskrit, and Punjab; the handicrafts for which the valley was celebrated were Central Asian; and the religious faith was variously Buddhist, Hindu, Sikh, and Muslim. Sufi masters left a legacy of music and tolerance in their Muslim teachings. A Sikh who had lived many years in Srinagar described the culture of the valley as an old cloth so covered in patches that you can't see the original.

Today, the valley is predominantly Muslim, but, as part of the maharaja's portmanteau state of Jammu and Kashmir, it still shares its destiny with other faiths and peoples: the Hindus of Jammu, the Buddhists of Ladakh, as well as Gilgits and Baltis, Hunzas and Mirpuris. There had been conflicts between the communities in the past, but by the mid-twentieth century Kashmir was an unusually tolerant culture. It escaped the intercommunal violence that Partition brought to the neighboring Punjab when the British left the subcontinent, in 1947. Kashmir's violence was to occur later, as the two new states of India and Pakistan became the latest of Kashmir's neighbors to fight over it.

Today, Kashmir is partitioned—Pakistan controls slightly less than a third, India some sixty per cent, and China the rest. Most of Kashmir's

twelve million people are concentrated in Indian-held territories, and the rest are mainly in Pakistan-held ones; relations among its many communities are now marked by mutual mistrust. And since the late eighties a bewildering number of combatants have fought a savage, irregular war that, in a steady daily toll of killing, has cost, depending on whom you believe, between thirty to eighty thousand lives. On the side of the Indian state, the participants include the local police, the Border Security Force, the Central Reserve Police Force, and the Army, supported by various intelligence organizations and a motley group of turncoat former militants who have muddied the public understanding of who, over the years, has done what to whom. Opposing them are a proliferation of Islamic militant groups. At one time, there were more than sixty of them. Several are fundamentalist and deadly—like the Lashkar-e-Taiba and Jaish-e-Mohammed, which are based in Pakistan (and have been listed as terrorists by the United States) and were recently banned by Pakistan's President Pervez Musharraf. The largest group, the Hizbul Mujahideen, is Muslim but not, its supporters insist, fundamentalist, and most of its activists, who number around a thousand, are Kashmiris.

Surrounding the insurgency is the wider, implacable hostility between India and Pakistan. But at its core is the story of a people who, for five centuries, have been longing to call their homeland their own.

Last October, I was permitted to go into what Pakistan calls Azad ("Free") Kashmir, a territory that Pakistan maintains is truly autonomous but which depends entirely on the country's military and money for its continued existence. India calls the territory Pakistan-occupied Kashmir. The entity has existed ever since Pakistan wrested this northwest third of the original state of Jammu and Kashmir from Indian

control in a war that followed the 1947 Partition. For Pakistan, that war was the first step toward a liberation of Kashmir's Muslims from India. Once liberated, Pakistan hoped, the Kashmiris would join Muslim Pakistan.

At the time of Partition, Jammu and Kashmir was still ruled by a Hindu maharaja, Hari Singh, a descendant of Gulab Singh. The maharaja was one of five hundred and sixty-two fabulously rich feudal monarchs whom the British had manipulated in order to maintain their grip on much of India. At Partition, these states were given a choice of joining India or Pakistan. Independence was not on offer. Most joined India. The maharaja dithered for months, unable to decide between two equally unattractive options. As a Hindu, he did not like Pakistan. As an Indian, he did not like the British. As a prince, he cared neither for the antifeudal Mahatma Gandhi nor for the local Muslim leader, Sheikh Abdullah, who favored autonomy for Kashmir but without its maharaja. Then, on October 20, 1947, armed tribesmen and regular troops from Pakistan invaded Kashmir. The maharaja appealed to India for support and hastily agreed to sign the now famous Instrument of Accession to India: the state of Kashmir and Jammu was accepted as part of the new federal union of India; in exchange, it was, exceptionally, granted a semiautonomous status. (India would control only matters of defense, foreign affairs, and communications; everything else was to be run by Jammu and Kashmir's own parliament.) Pakistan, furious, refused to accept the legality of the accession, and Pakistan and India fought their first war over Kashmir.

In Pakistan, what is remembered was a promise made by the Indian Prime Minister Jawaharlal Nehru to hold a plebiscite in which the people of Kashmir could make their preferences clear. That plebiscite was never held. India blames Pakistan: in 1949, after a ceasefire was agreed to under United Nations supervision,

Pakistan failed to withdraw from Azad Kashmir, a betrayal that, India says, vitiated the commitment to the plebiscite.

Today, there are few routes that connect Azad Kashmir with Pakistan proper. Some fellow-journalists and I set out from Islamabad at 6:30 A.M. and drove for five hours along vertiginous valleys, through Muzaffarabad, the capital of Azad Kashmir, and on into the mountains to Chakothi, a town on what is now known as the Line of Control—the ceasefire line established in 1949, after that first war over Kashmir. There, we walked to a peaceful clearing and sat sipping fruit juice. An immaculately turned-out brigadier, Mohammed Yaqub, the commander of the sector, briefed us on the Pakistani version of the history of the present conflict.

Tensions were unusually high. The United States bombing of Afghanistan had begun, and the military's view was that India might take advantage of the situation—troop movements had been detected. Yaqub's list of the casualties incurred in the last thirteen years of what he saw as Kashmir's freedom struggle against India was startling, even if undoubtedly exaggerated: 74,625 killed, 80,317 wounded, 492 adults burned alive, 875 schoolchildren burned alive, 15,812 raped, 6,572 sexually incapacitated, 37,030 disabled, 96,752 missing.

We took a path that led to a bluff overlooking a tributary of the Jhelum River. There was a slender, deserted bridge. On the other side were the Indian Army fortifications. A line of washing flapped in a light breeze above a series of bunkers. I peered through binoculars at men peering through binoculars at me. They waved, I waved back. A Pakistani officer admitted that, in more relaxed times, he met his Indian counterparts on the bridge and shared tea and sweets. "We don't talk about the war," he said.

Just as night was falling, we stopped at a refugee camp about an hour's drive away. A camp manager called on the refugees to tell stories of the atrocities that had forced them from their homes in Indian-controlled Kashmir. The misery, no doubt, was real, but the exercise smelled too much of propaganda to be of any genuine interest. The message, though, was clear: Kashmir was the unspoken subtext of the Afghan war. Under President Musharraf, Pakistan has sided with the United States and backed the bombing of Afghanistan. Nearly twenty years earlier, Pakistan had also sided with the United States in its mission to end the Soviet occupation of Afghanistan, by enlisting Pakistan's Inter-Services Intelligence, the I.S.I., to arm and train Islamic warriors to lead the fight. The I.S.I. had seen the opportunity to foment discontent in Kashmir, and Islamic warriors were armed and trained and sent there as well. Both wars were seen as religious and patriotic causes. But now Musharraf had renounced the Taliban and his country's earlier ambition to dominate Afghanistan through support of its hard-line Islamist government. Would he also be forced to abandon a dream that Pakistan has clung to since 1947—of uniting the Muslims of Kashmir with the state of Pakistan?

I met a member of one of the Kashmiri militant groups in Islamabad. He called himself Iqbal, though we both knew that it was not his name. He was a good-looking man in his early forties, with black hair beginning to gray. We had arranged to meet in an outdoor café. He was nervous, and constantly scanned the customers until he insisted that we move to a different location. We drove around the city looking for somewhere to talk. Eventually, he took me to a house in an affluent district of the city, a two-story villa set back from the street by high walls. There, we

sat on the floor, and he told me his story.

Iqbal had grown up in a Kashmir that preserved the memory—from before the Moghuls—of an independent country. For him, the Instrument of Accession was important because, in granting special autonomy, it implicitly acknowledged the idea of Kashmiri independence. But the Indian government, anxious about Pakistan's ambitions and uncertain of Kashmiri loyalty, regularly encroached on that autonomy. In 1953, Kashmir's popular Prime Minister, Sheikh Abdullah, was removed and arrested (he was suspected of autonomous leanings)—the first in a series of detentions that continued through the sixties. In 1963, a sacred relic—a hair of the Prophet's beard—disappeared from the Hazratbal mosque in Srinagar, and demonstrations erupted. The following year, India passed an order that allowed the Indian President to rule directly in Kashmiri affairs. By then, Muslim sentiments in the valley were hardening.

The long-established Kashmiri tradition of tolerance—the pluralism that had accommodated so many different faiths and cultures—was breaking down in the frustration generated by India's interference. One friend described to me what the valley was like in the late seventies and early eighties. There were, he recalled, fevered political discussions, stimulated by activist teachers who distributed everything from the works of the Egyptian Muslim Brotherhood to the teachings of Mao Zedong. Kashmiris were impatient for change. Their hopes were focussed on elections that were to take place in 1987.

The chief minister was Farooq Abdullah, who had returned to power after having been dismissed by Indira Gandhi, in 1984. He had regained his position by allying himself with the Indian National Congress Party, and many regarded him as a traitor to the cause of Kashmir. The opposition was led by the Muslim United Front, a coalition of

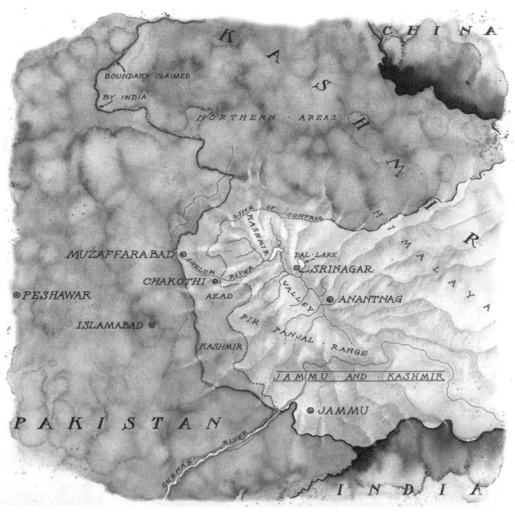

Mike Reagan

Kashmir has been occupied by Moghuls, Afghans, Sikhs, and the British, who sold it.

ten or so Muslim parties campaigning on a platform of Islamic unity and greater autonomy for Kashmir.

Iqbal was a student at the university in Srinagar and was involved in the electoral campaign. "The 1987 elections were our last hope," he told me. Eighty per cent of the population of the valley turned out to vote. When the results were declared, Farooq Abdullah's pro-India Conference-Congress alliance had—to the dismay and disbelief of the voters— won a two-thirds majority.

The fraud had been crude and blatant. In one constituency in Srinagar, witnesses told me, the result had been publicly declared, only to be reversed an hour later. After the elec-

tion, opposition candidates and party members were arrested. There were widespread street protests, which were brutally suppressed. "When the results were declared," Iqbal said, "people decided that we could not free our land through peaceful means."

Iqbal joined an underground group and was arrested. "I was in jail for two and a half years without trial," he said. When he was released, he was immediately rearrested and held for another two years. He was released and arrested again.

Iqbal returned to the university in Srinagar, but during an Army search an informer identified him as a mili-

tant. He was detained again. This time, he said, he was tortured. (According to Amnesty International, in Jammu and Kashmir torture has become so routine in the arrest-and-detention process that it is rarely reported.) But this time, once he was released, Iqbal took up arms. He joined Hizbul Mujahideen and spent three years as an underground militant. He was arrested three more times, before he finally escaped to Pakistan. He had spent, he said, fourteen years in prison, about half his adult life. "If you want to talk about Kashmir," he said, "you must talk about the eighty thousand innocent martyrs. It's a death rate of fifteen innocent civilians to every one Indian soldier."

A few weeks after I met with Iqbal, that balance shifted, marginally but dramatically, in the other direction. At eleven-forty on the morning of December 13th, in New Delhi, five men dressed in olive-green fatigues and armed with assault rifles, grenades, and explosives drove a white official car, complete with flashing lights and security passes, through the gates of the Indian Parliament complex. The session had just ended, and the politicians were beginning to disperse. It was only after security guards noticed the car turn the wrong way that they became suspicious. A guard ran after it, calling to the driver to stop. Alarmed, the Vice-President's security guard, waiting by his official vehicle, challenged the white car. Gunfire came from it as the car rammed into the Vice-President's vehicle, and the men inside ran toward the Parliament building. In the ensuing firefight, all five terrorists were killed, along with eight security personnel and a gardener. The car was found to be packed with explosives. The target of the assault was the Parliament building itself. Although the identity of the terrorists was not established, the Pakistan-based groups Lashkar-e-Taiba and Jaish-e-Mohammed were named by the Indian government, and the Indian press published calls to finish this long quarrel with Pakistan once and for all with a full-scale war.

In Pakistan, President Musharraf condemned the attack and banned the two groups, closing down their offices and arresting dozens of their members as well as other extremists. But in the following weeks tensions between the subcontinent's hostile neighbors heightened, and India began to lay mines along the border. The future of Kashmir was once again reduced to a poisonous contest between these rival nations. Each claimed to have the loyalty of the Kashmiri people and blamed the other for the conflict. In this deafen-

ing exchange, the voice of Kashmir was silent.

2.

Ramesh Mahanoori is a Pandit, a Kashmiri Hindu who, like most of the half million people in his community, lived until 1990 in relative prosperity in the Kashmir Valley. The Pandits formed the backbone of the professional class, and filled most of the teaching and government jobs. They had deep roots and high status in the valley, and lived side by side with Muslims, sharing the Kashmiri traditions of song and poetry, eating in each other's houses, sometimes worshipping at common holy sites. The harmony between the communities, so distinct from the tensions and violence elsewhere in India, was part of a general culture—the so-called Kashmiriyat—of which both Hindus and Muslims were proud. Even when Partition unleashed mass murder between Muslims and Hindus elsewhere, in Kashmir neighbors of different faiths preserved their courtesies and communal tolerance.

But between January and March of 1990 that tolerance ended, and a quarter of a million Pandits fled the Kashmir Valley, driven out by murders, riots, and death threats. The Pandits had become early victims of the new Muslim insurgency.

Ramesh Mahanoori was once a teacher in the Kashmir Valley. Now in his fifties, he received me in a tiny one-room house in a refugee camp on the outskirts of the city of Jammu. A large bed took up much of the room. There was a sliver of living space where a child sat on the floor, bent over a book; behind a curtain was a crude kitchen where Mr. Mahanoori's wife could be heard preparing tea. A tap outside served as the bathroom. We sat, cross-legged, on the bed, along with two of Mr. Mahanoori's friends. As we talked, his wife appeared and burrowed beneath the bed.

"That's where we keep the stores," he explained.

For Mr. Mahanoori, the expulsion of the Pandits was a straightforward case of betrayal. It began, he believed, with the Islamist underground, financed by the I.S.I. Its leaders had started organizing in Kashmir in 1986, and after the farce of the 1987 elections their followers increased. In 1989, an orchestrated campaign of executions began. "The first assassination was of a lawyer," Mr. Mahanoori said. "It was followed by other killings—three hundred and ninety highly selective killings of doctors, engineers, educators, judges. All labelled Indian agents. All our intellectuals." The government, he said, gave no protection. "There was a clear message from the majority that they could no longer live with the Pandits. The Muslims were all united under the banner of *azad*—freedom. Pakistan was their mentor."

Warnings were posted that Pandits who remained in the valley would be killed. Muslim activists set businesses on fire as the police stood by. Fear gave way to panic, and families began to leave. There were rumors of death lists in the local mosques. Pandits throughout the valley hastily packed and fled. Many ended up in the Hindu-dominated security of Jammu, in the south, imagining that they would return in a few months. But after they left their property was looted. Twelve years later, most of them are still refugees.

I heard endless variations of the Pandit story. Some people believed, genuinely, that the assassinated men had been agents of the Indian state. Others believed that the violence had been orchestrated from New Delhi (thus the lack of official interference); this way, the Kashmiri insurgents could be condemned for ethnic cleansing and dealt with accordingly. All versions agreed that the expulsion was brutal, sudden, and comprehensive.

Mr. Mahanoori recalled that in the village where he grew up in Kashmir he had been surrounded by members of his extended family. In the flight, they have scattered, and they rarely meet. "We had the same surnames as the Muslims," he said. "We were all related. They just converted to Islam—only three hundred years ago. Our cultures resembled each other. Here, in Jammu, we are aliens. We have nothing in common with these people."

Now there is a generation of children growing up in a world bounded by the camps. For them, Kashmir is a name, the source of their parents' sadness, something that marks them as different from the people of Jammu. They no longer speak the Kashmiri language, Mr. Mahanoori said. He longed for war. India, he said, should go to war with Pakistan, to resolve this issue once and for all.

3.

As I flew to Srinagar, I had few fellow-passengers—some Indian military personnel, a handful of Kashmiris, and one other foreigner. Even the most intrepid trekkers now prefer to explore other, less dangerous mountains, and the tourist trade that used to sustain the economy has dwindled.

The road into the city was an obstacle course made from an eclectic selection of barriers: metal bars set with eight-inch-long spikes, rolls of razor wire, and oil drums filled with concrete, which forced cars to weave a slow slalom path between them. Each barrier was guarded by men with automatic weapons. Beside one, in a bizarre juxtaposition, a poster offered a seductive welcome. "Kashmir—an adventure," it said. "The land of forests."

The light was fading as I reached the last barrier before my hotel, which had been recommended as a secure place to stay. The driver stopped and switched on the interior light. Beyond my own reflection in the glass, an armed guard was peering suspiciously into the car. In the deserted lobby, Muzak was playing

to empty armchairs. Three men looked up from the reception desk in surprise. The lobby was so cold that I could see their breath in the dank atmosphere.

India has now fought three direct wars in Pakistan, two of them over Kashmir. For India, the insurgency that began in the late eighties is another war with Pakistan—a proxy war, in which the enemy is Pakistani-trained infiltrators, with weapons and money supplied by the Pakistani intelligence services. This invasion of its territory, India argues, is a straightforward attack on its sovereignty, and demands defending. For India, Kashmir's status is incontrovertible: it is a part of the Indian state, a senior official told me, and there is no negotiation on either sovereignty or territory. India's response, therefore, has been a military one. But the nature of that response has created a conflict with the wider population of the Kashmir Valley.

Early in the morning on January 13th, four days before I arrived in Srinagar, two men were shot dead by Indian security forces, on a road near Dal Lake. Their names were Ahmed el Bakiouli and Khalid ed Hassnoui, and it was reported that they were foreigners who had attacked a Border Security Force patrol. In the ensuing incident, the two men were fired on by soldiers on watch in a fortified bunker nearby. By the time local photographers arrived, Ahmed and Khalid were dead.

The fact that Indian soldiers had shot two men was not in itself newsworthy. Since 1947, India has maintained a heavy security presence in Kashmir, one that is now half a million strong. To the local Kashmiris, these forces look and behave like an occupying army. With the exception of the local police—who are regarded with suspicion by the Army and the paramilitary forces drawn from elsewhere in India—few of these forces speak Kashmiri. They, in turn, are far from home, surrounded by people whose language they can-

not understand, and threatened by an enemy they cannot identify. To the Indian security forces, anyone they encounter could be a terrorist infiltrated from Pakistan. "If a dog barks in the market," one trader told me, "the Indians call him a Pakistani."

The men the solders are looking for belong to any number of dangerous militant groups, many with competing objectives—some wanting independence from India, or an Islamic state, or a union with Pakistan. Several groups began to impose a more severe version of Islam on the tolerant culture of the valley: women were made to wear veils, and bars and beauty parlors were closed down. Foreigners were attacked. In 1995, a Pakistan-based Islamic rebel group kidnapped six Western trekkers: four vanished and one escaped; the sixth was decapitated. There were plane hijackings, which sometimes led to the release of captured terrorist leaders; armed encounters in the mountains and villages; and car bombs and grenade attacks in the cities.

Indian security forces responded with repressive tactics. Shopkeepers and university professors, impoverished farmers and well-heeled businessmen continue to complain of routine cruelty exercised by the security forces during cordon searches: entire districts are sealed off, and the inhabitants are turned out of their houses and made to squat in the cold for hours as the troops ransack their homes. Men and boys are beaten; there are shootings; valuables go missing.

> "We are keeping them safe," an Indian officer said of the Kashmiris. "You never know which vehicle a terrorist might be driving."

For the Indian security forces, such operations are a necessary part of a war against an unseen enemy—

one who might be disguised as a market trader or as a schoolboy or even, as in the case of Ahmed and Khalid, as a pair of out-of-season travellers. But this time it was not just the people of Srinagar who were skeptical of the official account. Ahmed and Khalid, it emerged, were neither Kashmiri nor Pakistani. They were Dutch nationals of Moroccan descent who had ostensibly come to Srinagar as downmarket tourists. They had valid travel documents, had signed in at the Foreign Registration Office in Srinagar, and had been spotted at the Tourist Reception Center by a rickshaw driver named Amin Bakto, who was there looking for business. Bakto had invited them to stay at his houseboat, and they had been there for a week, when, according to an inspector general of the Border Security Force, they had gravely injured two of his men in an unprovoked terrorist attack.

Amin Bakto's houseboat, the Happy New Year, sits in a dirty side canal, greasy green water lapping against the boat's peeling paint. Bakto is a small, spare man, and he talked in nervous bursts, as though he were unable to shake the apprehension that he might somehow be implicated in the events that had led to the deaths of his paying guests. He lives with his family on an adjacent houseboat, and was willing to show me where the two Dutch nationals had slept for the week that they had been his guests.

It was a small room that contained little more than a double bed, which they had shared. Gaping holes in the floor were the consequence of the police search that had followed the killing. The room smelled of stagnant canal. Ahmed and Khalid had paid him two hundred rupees a night (about four dollars), Mr. Bakto told me, a sum that included the use of a heater and breakfast, which he had served himself at nine o'clock each morning. The men were pleasant and quiet, he said, and occasion-

ally played with his children. He never saw them pray or visit the local mosque.

On the day they died, he had gone to offer them breakfast as usual, but found that they had left. The Happy New Year was empty, the doors and windows open. The men had set out along the towpath to a nearby road. By 7:20 A.M., they were both dead, sprawled some twenty yards apart on a road now spattered with their blood. Later, the police had found on the houseboat the packaging to a pair of large kitchen knives, apparently bought in a local bazaar. The knives, in the police version, were the evidence that connected the two Dutchmen to a network of international terror.

No local witnesses came forward to corroborate the security forces' story, and almost nobody I met believed the account. The version favored by the local newspapers was that the patrol had been abusing a local woman, and the two Dutchmen had attempted to intervene. Others believed that they had been challenged by the patrol on their way back from morning prayers. They might not have understood an Indian soldier's command to halt. In either case, they risked being shot.

I went to visit Inspector General Gill, who commands the Border Security Force in Srinagar, and whose men had killed the Dutchmen. I had met him on my first evening in town at a rather stiff party attended by the local commanders of the security and intelligence forces in the district. He had seemed cultured and courteous, and it was difficult to connect him with the acts of torture and repression blamed on the men he commanded.

He had suggested that we meet at his bungalow in a hilltop compound that houses government servants. At the first barrier, my car was searched, and the driver and I were body-searched. At the second barrier, we were assigned guards to take us through the third barrier, where an armored car was parked, guarding the approach road. The final bar-

rier was beside the compound gate. From there, I walked to the house. We talked in a small, bare sitting room, warmed by a large metal stove that crackled in the corner.

Gill is a slim man of fifty-one, a Sikh from the Punjab. For him, there was no doubt that the two Dutchmen were terrorists. Their attack, he said, had been unprovoked. They had inflicted eight stab wounds on his men before they were shot; one of his men lost an eye. For Gill, the Dutchmen reinforced his conviction that the war in Kashmir was sustained from outside—a Pakistani proxy war. He admitted that there was no evidence of a Pakistani connection in the Dutchmen's case. "Do I have to prove that everyone has a past career?" he asked me plaintively. He held to his general point: If Pakistan, with its connections to international terror, would stop sending militants into Kashmir, the trouble would subside overnight. It was a conviction that was shared by the Indian government and widely reflected in India's national press. Besides, he insisted, his men did not shoot people without cause, and the many allegations of torture and disappearance made against his forces were scrupulously investigated. And almost none, he said, stood up.

The local press was unconvinced, even though it had been thoroughly briefed on the incident by Gill himself. The widely held feeling in the valley is that the insurgency is no longer masterminded from Pakistan or anywhere else: the native-born movement is now well established, after years of Indian abuses. And that feeling was reinforced by the killing of the Dutch tourists, regardless of what actually happened. Perhaps the men, armed only with kitchen knives, attacked a military patrol. But the belief is that this army of occupation can shoot anyone it wants to, anytime, with impunity.

I tried to explore the region around Srinagar. The roads to the border, where Indian and Pakistani troops continued to exchange mortar fire, were blocked with snow. To travel outside the city was dangerous. The splendid Moghul fort perched on a hill above Srinagar was occupied by the Army. I found myself circling the city, trying not to feel caged. The streets were wet and muddy, with piles of dirty snow. The light was flat and weak, filtered through a morning fog that rarely dispersed during the day. As I drove around, the sense of military occupation was oppressive. On every street, people were being stopped and searched by Indian soldiers, taxi-drivers opening the trunks of their cars for inspection, lines of bus passengers waiting to be frisked. My car was frequently stopped, my documents inspected, and my driver closely questioned. I was harangued by Indian soldiers who considered the stamp on my press pass insufficiently clear.

In the evenings, Srinagar was subdued, and the tension was unmistakable. As night fell, the people I met and talked to often began to fidget, caught between the obligations of hospitality and their anxiety that I leave before the streets became unsafe. Nobody, they told me, goes out after dark. A tremulous sociology professor described to me the social effects of the long war—migration, unemployment, broken families, a startlingly high rate of suicide.

"It's the constant fear," he said. "Torture, tension. Even at home, the security forces can arrive at any minute. We used to be a leisured people. Now all our entertainment has gone. It's out of the question to go out."

Education had deteriorated in the wake of the Pandits' departure, he told me; young women cannot find husbands, married women are widowed and destitute. He urged me to walk around the old city, a district I had been warned against, to discover how people really felt. It was a hotbed of militancy, I was told, and subject to constant cordon searches.

"Talk to people," the professor said. "No one will harm you." After

a pause, he seemed to think better of his assurances. "Don't tell anyone in advance. Don't make an appointment, in case, in their innocence, they tell someone you are coming. And don't stay more than half an hour in the same place."

It was now dark, and he was agitated. I drove through the rapidly emptying streets, ready for another evening of chilly confinement in my hotel. But that evening I had a visitor.

I had called Commander Chauhan, of the Border Security Force, several times, and now he appeared, exuding friendly confidence, eager to show me the sights—at 9 P.M., long past the hour when civilians had abandoned the city. The Commander was a portly, bespectacled man dressed in a camouflage jacket and a black beret, and carrying a polished swagger stick and a walkie-talkie. He bustled jauntily into the hotel's freezing dining room and greeted the waiters by name. The waiters smiled anxiously.

The Commander was eager to stress that he had excellent relations with the local people. His job, he said, was to protect them from the militants. His unit had adopted a girl who had been attacked with acid by Islamic fundamentalists for failing to wear a veil. There were orphans whom his men took care of. And, he assured me, they were steadily weaning the Kashmiris off Islamic extremism. Normality, he announced, was visibly returning.

"You see girls driving cars, boys and girls on motor scooters, going out to the lake, going to hotels, cinemas, and beauty parlors," he said. I had seen none of those things.

The security forces, as he described them, were dedicated to social welfare: "If someone's wife goes into labor in the evening, they just ring up. I send a car to take her to hospital. I have had so many calls to say thank you."

I didn't doubt it. Anyone moving around the city at night without military protection, pregnant or not, risked being detained as a terrorist.

As though reading my thoughts, Commander Chauhan suddenly said, "What have you seen? I wish we had met earlier. I could have taken you on the lake. We have motor launches, you know. I'll take you out now, to see the city."

Outside, his jeep backed slowly to the door, as six soldiers armed with automatic weapons walked alongside. I climbed into the front, the Commander took the wheel, and the soldiers jumped into the back. Another vehicle moved up behind us. We then pulled out into the deserted street and embarked on a tour of the city.

"This is the polo ground," he said. He waved a hand vaguely at the darkness. "But they don't play much polo these days. And here—this is the golf course. Excellent golf." We drove on in the empty street. "This is the canal." We turned toward the lake. "Have you seen the Nishat Gardens?" he inquired.

Before I could reply, I was blinded by the beam of a searchlight which had appeared from inside a bunker. The Commander braked sharply. A man jumped out from the back of our vehicle and explained our presence to a group of nervous soldiers whose guns were trained on him and on us. Satisfied, they allowed us to pass, and the Commander continued with his description of the delights of boating on Dal Lake. I found myself recalling an incident described to me by a lecturer in the English department of the Srinagar university. A group of graduate students doing research on the lake one day were shot dead in their boat.

A few hundred yards further on, we found ourselves inching around oil drums and rows of spikes in the road, and came to a stop before a final improvised barrier of rocks—and another checkpoint. At the next bunker, however, we failed to stop in time, and there was a fusillade of hostile shouts. Commander Chauhan braked violently. A soldier in the back climbed out, his hands raised, and stood for several minutes in front, in the glare of the headlights, trying to talk down the guard whose gun was trained on him. I held my breath. Five more guns were pointing at the jeep. Even Commander Chauhan had fallen silent. At last, the soldier gradually lowered his arms.

Commander Chauhan had lost a little of his bounce. For the first time, he seemed to feel that he should acknowledge the surreal character of a city tour in which even a senior officer risked being shot by his men. "Actually," he said, "we give them orders that every vehicle must be stopped. You never know which vehicle a terrorist might be driving." As he picked up speed, his faith in his mission returned. "We are keeping them safe," he said. We turned back through the deserted streets of the old city. There was hardly a light showing, but for Commander Chauhan this didn't mean that people were afraid to go out. "Look," he said triumphantly, "they are all in bed with their wives and their blankets, and my men are out here, keeping them safe!" Suddenly, he turned to me. "What did people tell you, by the way?"

"They said they wanted independence," I replied bluntly. "That they were afraid of your men and their searches." I stopped short of telling him how many rejoiced when Indian soldiers were killed.

"Did you ask them why we search them?" he said.

I had not, of course, though I could imagine Chauhan doing so, chattering as he frisked people, in an exercise in hearts-and-minds didacticism, cheerfully explaining his motives as their humiliation deepened.

I asked him how many terrorists he thought there were.

"Very few, these days," he replied.

Why, then, did the government need to keep half a million men here?

"Because," he replied quietly, "you don't know who they are."

4.

The conflict over Kashmir has entrenched the worst suspicions that India and Pakistan nurse about each other. For India, the separate Muslim state of Pakistan represents a rejection of the secularism that India believes to be essential to keeping its own rival religious communities at peace. If Kashmir's Muslims were to join Pakistan, what signal would that send to the more than a hundred million Muslims elsewhere in India? For Pakistan, India's refusal to allow Kashmiri Muslims to join the Pakistani state merely confirms its conviction that India never abandoned a long-term ambition to establish Hindu domination on the subcontinent, or that it even accepted Pakistan's existence. But for the people of the Kashmir Valley, with their distant dreams of independence, neither neighbor offers a solution. Pakistan's muscular Islam is at odds with Kashmir's Sufi-inspired traditions. The Muslims and Pandits of the valley speak a different language from the language of India or Pakistan. Neither country is home, and each, in turn, has been a threat: after all, it was the incursion of tribal raiders from Pakistan in 1947 that brought Indian troops in retaliation.

Thirteen years into the insurgency, the local politicians in Kashmir, like the competing militant groups, have conflicting objectives. Even members of the All Parties Hurriyat Conference, which was formed in 1993 by more than thirty political parties to act as a voice of a people who felt themselves disenfranchised, are quarrelsome and deeply divided. I met many members, and asked them what they wanted for the country. I got many different answers. One wanted union with Pakistan. One wanted independence. Others would be happy with real autonomy within the Indian state. Some had links to the militants; others did not. Each claimed to represent a general majority.

From civilians I got a different picture. Weary of war, few believed that anything could be won, now, through the armed struggle. But few supported Gill's contention that the hostilities would end once Pakistan stopped supporting them. I understood this view when I met some of the young new recruits. There seem, potentially, to be an endless number: the war's capacity to create new militants is limitless.

Militants are buried in Srinagar's many martyrs' cemeteries, some of them large adjuncts to regular graveyards, others crammed into small corners across the city. They are crowded with almost identical concrete gravestones, covered in Arabic inscriptions in green lettering. A few bear English place names—"Birmingham" was one I noted—an indication of Kashmir's appeal to disaffected Western-born Muslims looking for a cause.

But exactly what that cause was, beyond the single word *azad*, was unclear. As I looked at the gravestones in one of the smaller cemeteries near Dal Gate one day, a group of boys gathered around me and laboriously translated the inscriptions. They were eager to show me significant graves—Islamic warriors from faraway countries, or men whose spectacular deaths had stuck in their memories. They pointed out professionals—lawyers, teachers doctors—and men who had died under torture. They called all the dead "martyrs," but they couldn't always tell me what they had died for—whether they were martyrs to Kashmiri independence, or to the union with Pakistan, or simply to Islam. One grave that was pointed out to me belonged to Aafaq Ahmed Shah, who, at the age of eighteen, had become briefly famous for inaugurating a new phase of the militant group Jaish-e-Mohammed's war, that of the suicide bomber. I had earlier visited his family in the old city.

His mother had opened the door and stood on the doorstep looking at me. She was a small, middle-aged woman with dark circles under her eyes, and she knew, before I explained, what I had come for. She remained immobile, tears flowing down her face, reluctant, it seemed either to turn me away or to admit me to what she knew would be a painful rehearsal of her grief.

Still sobbing, she let me in, and I sat on the floor of a freezing room, waiting for her husband to return from the market. Mohammed Yusuf Shah was a thin, elderly man, a retired college teacher dressed in a brown *pheran*. He settled beside me as his wife brought blankets and fire baskets and poured us cups of chai from a thermos flask.

"My son was nearly nineteen years old," Mohammed said. "He wanted to be a doctor. There's a photograph of him"—he waved his hand vaguely—"somewhere, wearing a stethoscope." He made no move to get it, as though already discouraged by the effort. His wife had begun to cry again.

"Mysterious are the ways of God," he said. There had been no warning that his son would join the militants. "He willed it. He did it. That is all. He was a good, silent, obedient boy. He was my son, but, more than that, he was my friend. He was here, dawn to dusk, every day, day and night."

On March 25, 2000, the boy disappeared. The family searched for him, fruitlessly. Three days later, he telephoned. "Father, I left," he said, and hung up.

On April 19th, dressed in an Army uniform and carrying an Army I.D., Aafaq tried to ram his car through a heavily fortified gate of the Army's XV Corps headquarters, near his home. The car exploded after a solider started firing at it. Five soldiers were injured; only one person was killed—Aafaq, who was blown to pieces. The family read of his death in the newspapers.

I found another family of a young martyr in a village some thirty miles outside Srinagar. We drove along long straight roads lined with tall poplars, past fields of saffron that were just showing a first flush of green shoots, past empty paddy fields, waiting to be planted. The village itself was along a muddy track, buried among trees, peaceful in the chilly morning. "Ignorance is the root of all evil" was carefully painted on the wall of the village school.

I sat on the floor of the family's small living room as villagers crowded in, competing to tell the story of Nazir Ahmed Khan.

"Nazir was in the tenth class," a young neighbor told me, and also wanted to be a doctor. "His hobbies were gardening, photography, and cricket."

Nazir's elder brother, Mohadin, drove a taxi to support the family. There was a skirmish in a nearby village, and the Army appeared at the door, convinced that Mohadin had driven a wounded terrorist to the hospital. Mohadin was not at home, but Nazir and his father were. They were interrogated, but the soldiers were not satisfied, and the father and son were both beaten, and then their limbs were held over a fire. Afterward, Nazir ran away. "He could not bear being tortured for no reason," the neighbor said. He had gone to join the militants.

Mohadin was summoned to the Army barracks, and he, too, was tortured and then imprisoned. The family sold the taxi to bribe the Army for his release; it was their only asset. And then Nazir was killed.

I went to see where he had died. We drove back to the main road, past a sign that read, "Our job is to make everybody see the beauty of Kashmir," then turned down a muddy track to the village. We inched along the narrow streets until a villager pointed to the house: the roof at one end had collapsed, and its supporting wall was a pile of rubble. I scrambled up the slippery lane and pushed open the ramshackle corrugated-iron gate. A small crowd followed me in.

The boy had joined two other militants, and the three of them, the villagers told me, were hiding in this house. An informer told the Army. The cordon search lasted for three days and three nights, and the entire village was made to squat in the cold on the recreation ground. Fire baskets and the old men's woollen hats were confiscated. Ten thousand soldiers came, I was told. I said that ten thousand soldiers is a very large number. The villagers insisted.

Cornered, the militants gave themselves away—one of them fired on the soldiers from an upper room—and the Army ordered seven villagers to walk up to the house and put two explosive devices inside. Everyone knew that the villagers would be harmed. It was, they said, a frequently used tactic. "The militants don't fire on civilians," a villager explained. "If you refuse to do it, the Army shoots you." The villagers got out before the devices were detonated. Nazir was eighteen, and had been a militant for a week.

Later this year, there will be elections in Kashmir—the opportunity, in principle, for the people to express their political will. But, after years of vote rigging and intermittent direct rule, Kashmiris have lost their faith in India's secular democracy. For the politicians in the loose coalition of the All Parties Hurriyat Conference, there is no point in even standing. To do so would require their swearing an oath of allegiance to the Indian state, which they do not wish to honor. And, at the very least, they want an autonomy that they believe India's current government—dominated by the Bharatiya Janata Party, an organization with an aggressively pro-Hindu ideology—will never grant. On February 12th, the Hurriyat announced that it would boycott the Indian elections and hold an election of its own—to choose representatives who will sit at a negotiating table with India and Pakistan. But India is not going to give up Kashmir, and the negotiations have no hope of succeeding.

President Musharraf, too, has called for negotiations, and on February 6th the U.N. Secretary-General, Kofi Annan, offered to mediate. For Musharraf, negotiations could be the key to his survival. He has declared his wish to make Pakistan a more secular state, attempting to dismantle the networks of Islamic extremists who, for the past twenty years, have systematically infiltrated Pakistan's government, Army, and, especially, its intelligence services, the I.S.I. These people are viscerally opposed to Musharraf's ambition. If he is to succeed—if, at the very least, he is to put an end to the I.S.I.'s support for cross-border Islamic terror—he needs to show that the cause of the indigenous Kashmiri struggle has not been abandoned. India, meanwhile, has not taken up the offers for negotiations.

In Kashmir, an end to the struggle seems ever more remote. Nazir, like Aafaq, had joined the ranks of the martyrs in a war that has lost its way, a war that now feeds on itself—each act of violence generating a new response that generates more recruits. For some of the valley's young men, it can seem as though there were little else to do—the war is their occupation. The Kashmiriyat is now a forlorn memory, and has been replaced by the cult of the gun. The people of the valley believe they are trapped in a war without end, in which anyone can become a victim. Tens of thousands have died. Scarcely a family in the valley is untouched.

From *The New Yorker*, March 11, 2002, pp. 64-75. Copyright © 2002 by Isabel Hilton. Reprinted by permission of the author.

The branding of Hong Kong

A dragon with core values

Stick that in your lapel

HONG KONG

ON HANDOVER day in 1997, Donald Tsang, then Hong Kong's finance secretary, pinned a little emblem on to his lapel. It was a double flag—Communist China's joined to Hong Kong's *bauhinia* flower—that stood for "one country, two systems". For more than three years Mr Tsang was rarely seen without it, and it became, along with his bow ties, his trademark.

Last year—by when he was chief secretary—Mr Tsang replaced the emblem with a little dragon, the fruit of three years of research by international brand consultants. Besides cosmopolitanism, says Kerry McGlynn, the government public-relations director behind the project, the dragon projects five "core values". These are three adjectives—"progressive", "free" and "stable"—and two nouns, "opportunity" and "quality".

The visual link, according to the government, is self-evident: Hong Kong stands for "East meets West". So the dragon is composed of two parts that could, if you twist it, stand for the letters H and K, as well as the Chinese characters *Heung* and *Gong*. Combined into a dragon, an ancient Chinese metaphor for energy, the strokes represent Hong Kong's legendary dynamism.

The dragon appeared on brochures, buses and much else last summer, and within days Hong Kong's people were naming it. Expatriates saw it mostly as a "flying fox", while Cantonese speakers—usually more creative in such matters—settled on "shocked chicken". Those appraised of the consultants' fees called it "the HK$9m dragon".

Perhaps the most perplexing thing about the dragon, however, is that it took Hong Kong so long to get one. Canada branded itself in 1970, and New York ("the Big Apple") a decade later. Besides, their fetish for brands is one of the few core values that most Hong Kong residents agree on. As one long-term resident puts it, "When the going gets tough, Hong Kong goes shopping."

Where Business Meets Geopolitics

A vast new pipeline that will bring Caspian oil westwards has opened, at a cost of $4 billion. Western countries see it as important in reducing their reliance on oil from Russia and the Gulf states. Like other pipelines in the region its route marks the shifting of political allegiances that has driven a wedge between Moscow and Washington

WHEN Tsar Nicholas I laid plans for a railway between Moscow and St Petersburg 150 years ago, he took the direct route. By laying a ruler on the map he decreed a straight connection between Russia's two great cities, bar a small kink where it is said that he accidentally drew around his finger—timid courtiers failed to alert him to his detour.

Today's most contentious oil and gas pipelines, mainly sited or planned in and around Russia, are not susceptible to such autocratic whim. On Wednesday May 25th, the 1,800km Baku-Tbilisi-Ceyhan (BTC) pipeline officially opened for business some 13 years after its conception and at a cost of $4 billion. The pipeline, built by a consortium led by Britain's BP, will bring Caspian oil from Azerbaijan across the Caucasus to the Mediterranean coast of Turkey, from where it will be tankered to markets worldwide. When it is fully operational it could transport 1m barrels a day, just over 1% of the world's current oil consumption.

The strategic value of Caspian oil did not escape Adolf Hitler. He fatally over-extended his army's supply lines with a dash to secure the region's oil reserves, resulting in a decisive defeat at Stalingrad. Russia kept control of the region's oil until the break-up of the Soviet Union. Then western governments and oil companies, searching for fresh sources of oil in a bid to reduce reliance on the Middle East, advanced on the Caspian themselves. But the region has failed to live up to its early promise. Azerbaijan is never likely to become the new Kuwait. America's Energy Information Administration estimates that the Caspian region has oil reserves of between 17 billion and 33 billion barrels, rather than the 200 billion touted in the mid-1990s, plus a fair bit of natural gas.

Despite this disappointment, more of this oil and gas is coming on tap and new means of getting it to market are required. Unlike many other big oil producers, such as Saudi Arabia and it neighbours clustered around the Per-

sian Gulf, Caspian oil and gas is landlocked and set apart from the sea by a selection of countries with differing ambitions and loyalties. Tsar Nicholas and his planning techniques have thus given way to complex geopolitical [maneuvering].

The quickest route to the sea is to go south through Iran. But handing control of a key pipeline to such an unpredictable regime was inconceivable. Instead, in 2001, a 1,510km pipeline (known as CPC) opened between the Tengiz oilfields of Kazakhstan and the Russian port of Novorossisk. This pipe, built by a consortium led by ChevronTexaco, was the first privately owned pipeline crossing Russian soil (though its government has a 24% stake). Russia, keen to retain control over oil exports, which it uses as a foreign-policy lever, has kept the rest of its 48,000km of oil pipelines strictly under the control of state-run Transneft.

The BTC pipeline, though the most expensive option for exporting Caspian oil, was backed by America because it avoided Russia, thereby reducing the dependence of the Caucasus and Central Asia on Russian pipelines. The pipeline also provided an opportunity to bolster regional economies that the West is courting, especially those of Georgia, Azerbaijan and Turkey, a NATO ally, and build support for America in the region. Georgia's location gives it a "strategic importance far beyond its size", according to America's State Department.

Upgrading an alternative route through Georgia to Supsa on the Black Sea would have made for a far shorter (and cheaper) pipeline. But Turkey complained that it would lead to an unsustainable level of shipping passing through the Bosporus Strait that bisects Istanbul. At Washington's urging, the BTC pipeline wended its complex way through Azerbaijan, Georgia and Turkey. However, some critics of the pipeline point out that the oil revenues provided to Azerbaijan will help to prop up the country's autocratic and corrupt regime. And environmentalists have complained that the pipe slices through a national park in Georgia.

Pipe hype

Oil-thirsty China is also keen to get its hands on the region's resources, and in September work began on a 1,000km pipeline from Atasu in central Kazakhstan to western China. Eventually, China hopes to extend the pipe another 2,000km across Kazakhstan to the Caspian oilfields. However, plans to link it with existing Russian pipelines to allow the movement of Siberian oil through Kazakhstan are likely to be stymied by Transneft.

In Europe, too, countries are not afraid to indulge in a little pipeline politics. In February, Ukraine gave Russia a slap in the face by agreeing to reverse the flow of the Odessa-Brody pipeline. This was supposed to take Russian oil south to the Ukrainian Black Sea port of Odessa and then on to world markets by tanker; it will now pump Caspian oil up through Ukraine and into the European pipeline network. Ukraine has recently experienced fuel shortages, which its prime minister blamed on over-reliance on Russian oil.

A little further afield, two other proposed pipelines demonstrate how the quest for energy resources can overcome some political difficulties while creating others. A thawing of relations between India and Pakistan prefig-

ured a recent announcement that India is considering a 2,775km gas pipeline link to Iran. America fiercely opposes involvement with Iran and wants to use outside access to Iranian energy resources as a lever to stop the country's nuclear programme. Russia has broadly supported Iran's nuclear programme, in part as a counter to America's ambitions in the Caspian region.

For its part, Pakistan is party to an agreement to build a gas pipeline from Turkmenistan, which has substantial gas fields, through Afghanistan to the Pakistani coast. The Taliban, still an active threat in Afghanistan, has vowed to disrupt the project if American firms are involved. But the threat of terrorism, even in these unstable parts of the world, is probably exaggerated. Although insurgents have disrupted exports of Iraqi oil by blowing up pipelines there, most pipes run underground and attacks on them have not been common. Other oil installations, such as refineries and depots, are generally thought to be at greater risk.

In fact, the vast majority of oil pipelines are uncontentious and unthreatened: there is 322,000km of plumbing carrying crude oil, gas and refined products in America alone, for instance. But as the race for oil and gas from remote areas hots up, threatening old alliances and forging new ones, the politics of pipeline placement will become ever more complicated. The days of the tsar and his ruler are over.

Oil Over Troubled Waters

A great game unfolds between America and Russia

IN CLASSICAL times, the Black Sea was perversely known as the *Euxeinos Pontos*, a sea friendly to strangers, even though its notoriously turbulent waters were nothing of the kind. The hope was that if you gave the place a nice name, the invisible powers who governed its towering waves might feel placated and behave more calmly. To this day, it remains a temperamental stretch of water that can generate sudden squalls and treat outsiders in unpredictable ways, even when efforts are being made to appease its restless spirits.

In 1992, the late Turkish president, Turgut Ozal, thought he could assuage those spirits for ever and turn the sea into a zone of peace and co-operation, where ancient trade routes would thrive anew. The fruit of that post-cold war vision is the Istanbul-based organisation for Black Sea Economic Co-operation. For over a decade, its members (all the littoral states, plus near neighbours Greece, Moldova, Albania, Armenia, Azerbaijan and, as of recently, Serbia) have trundled along to meetings without ever realising Mr Ozal's vision. The fact that Armenians and Azeris were locked in armed confrontation, backed respectively by Russia and Turkey, has hardly helped.

About a month ago, and entirely unnoticed by the world, BSEC suddenly did something rather unfriendly to a stranger. It flatly turned down a request from the United States for observer status. While the brush-off was explained in arcane procedural terms, it was an open secret that Russia had blocked the application—to the embarrassment of the group's other ex-communist members. In fact, eight of them issued a separate statement saying Uncle Sam's presence would have been a welcome boost, and they regretted his exclusion. (If NATO members Greece and Turkey had any feelings on the matter, they did not air them.)

What America would have done if it had attained its lofty ambition may never be known. But to judge by the word on the think-tank circuit, there is a strong feeling in Washington that the Black Sea region is ripe for transformation into a new sort of security club, whose members co-operate to keep ports and pipelines safe from terrorists and other undesirables.

As steadily increasing amounts of energy flow into, and out of, the Black Sea, the stakes are certainly high. This week saw the formal opening, in Azerbaijan, of one of the world's most important energy conduits, a 1,770-km (1,010-mile) oil pipeline linking Baku in Azerbaijan with the Turkish port of Ceyhan via the mountains of Georgia. Gas from Azerbaijan, Iran and possibly east of the Caspian will soon be flowing along a similar route into Turkey, and thence to southeastern Europe. The pipeline promises to bring a bonanza for Azerbaijan, and a modest boost to the hard-pressed finances of Georgia.

While America has taken the lead in lobbying for the construction of pipelines which bypass Russia, and therefore deny the Russians any chance to use energy as a political weapon, it is the European consumer who will be most affected by these emerging routes. On present trends, Europe's reliance on Russian energy will increase sharply, whatever happens; the new pipelines will ease that dependence.

But a complex pattern of interests is already emerging. A recently constructed gas pipeline has started bringing energy across the Black Sea from Russia to Turkey. That has reinforced a burgeoning economic relationship between those two historic competitors and made it harder for the Turks to side unequivocally with the Americans if the contest for influence in the Black Sea ever becomes a straight fight between America and Russia. Indeed one school of thought in Washington regards the "old NATO" partners, Turkey and Greece, as less reliable than the eagerly pro-American countries that have only recently emerged from the grip of communism, and are poor and vulnerable enough to be grateful for anything they get.

One reason for heightened American attention to the region is the sense that the future of many countries is still a wide-open question: they could follow Central Europe into the warm embrace of western institutions or they could slide back into authoritarianism or stagnation. Bruce Jackson, an influential American lobbyist for NATO's expansion, put the point dramatically in some congressional testimony in March: "The democracies of the Black Sea lie on the knife-edge of history which separates the politics of 19th-century imperialism from European modernity."

The very fact that some parts of the region are quite advanced on the road to "European modernity" could be a divisive factor. One of the BSEC's more effective bits is its financial arm, the Black Sea Trade and Development Bank, which issues credits for export finance and cross-border projects. Its strategy director, Panayotis Gavras, says much the biggest factor driving investment in the region is proximity to the European Union; investors look eagerly at Bulgaria and Romania, which stand on the Union's threshold, and view other places far more warily.

As Britain prepares to take over the EU's rotating presidency, many people are expecting a fresh Black Sea initiative: something that would give heart to countries doing "well" in western eyes without dashing the hopes of the laggards and, if possible, without alienating Russia.

As Foreign Office mandarins ponder their options, they can take heart from some of the region's pleasant surprises. On June 6th, BSEC members will gather in Yerevan, the capital of Armenia, for a meeting of their affiliate bank. According to Turkish data, trade between Armenia and Turkey is precisely zero; the border is sealed, out of solidarity with Azerbaijan. As the delegates will observe, every shop in Yerevan brims with Turkish goods.

Central Washington's Emerging Hispanic Landscape

Scott Brady

Central Washington's Recent Transformation

In 2000, approximately 35 million individuals, or 12%, of the U.S. population classified themselves as Hispanic (U.S. Department of Commerce, Bureau of the Census 2000). That number exceeds the Census Bureau's 1992 middle series projections for this group's 2000 population, and reflects Hispanics' growing importance in the United States (U.S. Department of Commerce, Bureau of the Census 1993). Most of the Hispanics (66.1%) are of Mexican ancestry. Almost one-half of Hispanics (44.7%) reside in the western U.S. (Therrien and Ramirez 2000). Historically, Hispanics of Mexican ancestry were geographically concentrated in the border states of Texas, New Mexico, Arizona, and California. However, during the past three decades Hispanics of Mexican ancestry have spread beyond the border region and comprise significant proportions of the populations of states as far-flung as Illinois, North Carolina, and Washington.

In the 1940s, two federal government projects initiated a transformation of the cultural landscape of Central Washington. The Columbia Basin Project (1946-1966) harnessed the waters of the Columbia River for hydroelectric power generation and, beginning in 1951, irrigation of more than 500,000 acres of shrub-steppe lands in the region (White 1995). Irrigation ushered in large scale, intensive fruit, hops, mint, onions and hay production in an area that formerly supported only dry land wheat farming and ranching. As a result, the area's cultural landscape has been altered dramatically during the last fifty years.

Prior to the beginning of the Columbia Basin Project, the U.S. federal government negotiated an agreement with the government of Mexico that promoted the annual legal migration of seasonal agricultural laborers from Mexico to important agricultural regions of the western United States. This agreement, known as the Bracero Program (1942-1964), substituted Mexicans as seasonal workers for resident farm laborers who had joined the war effort during the Second World War (Gamboa 1990).

The agricultural changes wrought by the Columbia Basin Project and the associated influx of large numbers of Mexican laborers under the Bracero Program led to significant Hispanic settlement in Central Washington. During the past 40 years Hispanics have exerted a growing influence on the region's demography and landscape.

Setting

The nine counties of Central Washington are framed by the Cascade Range to the west, the Okanogan Highlands to the north, the Palouse to the east, and the Horse Heaven Hills and Blue Mountains to the south. Excluding the forested foothills of Chelan, Kittitas, and Yakima counties, the mean annual precipitation of this region is less than 10 inches. Under natural conditions, shrub-steppe covered this interior basin.[1] The Columbia River enters the region at its northeast margin and flows westward along the base of the Okanogan Highlands before turning south and flowing through a deeply incised channel cut by the ancient Missoula Floods. The river then turns southeast and flows to its confluence with the Snake River at the region's southeast edge.

During the region's homesteading era (1860-1910), settlers engaged in free-range cattle and sheep operations and dry land wheat farming (Meinig 1968). These economies persisted during the planning and construction phases of the Columbia Basin Project (CBP), which extended from the 1920s until after World War II. The Columbia Basin Project's irrigation phase began in 1951; by 1966, the CBP had delivered water to 500,000 acres of the basin. Irrigation of the basin led to the emergence of the region's current fruit, mint, hops and hay production economy. Since the initiation and completion of the CBP, Central Washington's population has more than tripled in size. In 1930 the region held 180,000 persons; by 2000, the region's population had grown to approximately 648,000. The decade of the 1950s had the most rapid population increase, when the population grew by 52%. This growth accompanied an expansion of acreage irrigated by the CBP. Growth in the population of urban areas on the west side of the Cascades (Seattle, Olympia, etc.) has matched that of Central Washington. Consequently, since 1930 Central Washington has supported approximately 11% of the state's total population.

Growth of the Hispanic Population

During the 1980s, Central Washington's white population decreased while the region's total population continued to increase. The increase resulted from the growth of the region's Hispanic population, the bulk of whom are of Mexican descent. Hispanic presence in the region can be traced back to the 1860s when small numbers of Mexican vaqueras and mule drivers serviced the ranching economy (Maldonado 1998; Watt 1978). Between 1900 and 1930, small numbers of Hispanic farm laborers from Mexico and California moved to the region, drawn by job opportunities in pre-CBR irrigated agriculture (Gamio 1971). World War II brought additional Hispanics as GIs from the Southwest reported to military installations in Washington, such as Larsen AFB in Moses Lake and the Ephrata Air Terminal (Maldonado 1998). However, not until the 1960s did Central Washington become what Maldonado (1998, 10) has called Hispanics' "Nuestra Hogar del Norte" (our northern home). The size of Central Washington's Hispanic population became increasingly significant following 1960.[2] Between 1960 and 2000, the Hispanic population grew 2,557%, from 7,596 people to 201,820 people. During the same period, the white population grew only 21%, from 360,702 people to 438,135 people.

Two factors contributed to growth in the Hispanic proportion of the population. First, the increase in irrigated acreage increased the demand for agricultural laborers. Second, the population structures of the white and Hispanic populations differ significantly. The Hispanic population is younger than the white population. The relative youth of the Hispanic population contributed to that population's more rapid increase between 1970 and 1990; the data suggests that, in the near future, the Hispanic population will continue to grow more rapidly than the white population.

Growth in the Hispanic population between 1960 and 2000 has changed county level demographics. In 1960, only Yakima County, the most populous in Central Washington, had a significant proportion of Hispanics. In 2000, however, Hispanics comprised more than 45% of the population of Adams and Franklin counties, and more than 30% of Grant and Yakima counties. Only Lincoln and Kittitas counties, where irrigated production of diversified crops remains insignificant, do not contain significant proportions of Hispanics. These data illustrate the rise of Central Washington's Hispanic population that occurred as irrigated agriculture expanded throughout most of the region.

Hispanic Places and Landscapes

A larger scale analysis of Central Washington allows the identification of Hispanic places. In 2000, 23 settlements with populations of greater than 1,000 persons were at least 20% Hispanic. Haverluk's (1998) study of spatial assimilation of Hispanic communities in the western United States included four Central Washington settlements (Moses Lake, Pasco, Sunnyside, and Yakima) and classified them as "new Hispanic communities." In new communities, Hispanics are recent migrants to Anglo-settled towns. Hispanics are underrepresented in the political process and there is greater pressure for them to assimilate in these dominant Anglo communities than in communities with longer periods of Hispanic presence or larger percentages of Hispanic population.

Despite their relatively recent arrival, Hispanics comprise the majority of the population in 13 of these settlements, and these places display the clearest traces of Central Washington's evolving Hispanic landscape. These settlements share several characteristics that give their landscapes a particularly Hispanic flavor, including: Spanish language churches; services provided by the state and federal government for Spanish language speakers; stores and eateries that cater to Hispanics; bus companies that offer transportation between Mexico and Central Washington; Spanish language newspapers; and leisure activities organized exclusively by and for Hispanics.

The emergence of Central Washington's Hispanic places has created a settlement pattern that mirrors colonial Mexico's agricultural landscape. Arreola and Curtis (1993) identified parallels between Mexico's colonial cities and its urban settlements near the United States—Mexico border in order to understand recent urbanization along the border. They noted the persistent compactness of border communities where the central plaza remained the core of activity. In the distinct periphery of border settlements, Arreola and Curtis found close-knit barrio communities served by family-run, neighborhood commercial establishments.

The model pertinent to Central Washington's Hispanic settlements is that of the colonial hacienda. Central Washington's Hispanic places were originally established and inhabited by Anglos to service the farms that surrounded them. During the past 40 years the Hispanic population in these towns has increased, while the Anglo population has decreased. Subsequently, much of Central Washington exhibits a spatial distribution of Hispanic and Anglos in which large tracts of Anglo-owned agricultural land surround towns inhabited by permanent and seasonal Hispanic residents. This pattern, with its large agricultural units inhabited by dispersed, Anglo landowners surrounding concentrated settlements of Hispanic agricultural laborers and small business owners resembles the *hacienda* of colonial Mexico. Similar to the Anglo-owned farms that surround central

Washington's Hispanic places, the *hacienda* lands were owned by *criollos*, a cultural group distinct from the laborers (Chevalier 1970). Similar to Central Washington's Hispanic places, a *hacienda's* laborers, Indians or *mestizos*, were concentrated in settlements, often called *cascos*, imbedded within the large agricultural units (West and Augelli 1989).

Hispanics have created a cultural landscape in Central Washington's small towns that is distinct from that of the region's large landowners. Hispanics have inhabited the town grids and dwellings of Central Washington and introduced traits from Mexico. The result is an archipelago of Hispanic settlements, in which Spanish is the first language and Mexican lifeways predominate, surrounded by a vast sea of irrigated agriculture where Hispanics work.

Like Mexico's *cascos*, Central Washington's Hispanic places contain houses of worship. Competing Protestant and Catholic churches in towns such as Mattawa or Royal City echo the religious rivalry that has emerged in Latin America during the past few decades. Hispanic congregations often inhabit church buildings that formerly housed Anglo Protestants. Hispanics also have built grand houses of worship that reflect the population's growing economic status.

Besides structures that meet Hispanics' spiritual needs, Central Washington places also have health and education services, including public schools that offer bilingual or Spanish instruction, health clinics, and doctor's offices that advertise *"se habla español."* In Othello, Hispanics purchased an abandoned church where they now offer meals to Hispanic seniors.

Hispanic-owned clothing and food stores, *carnicerias* [butcher shops], and restaurants compete for space on Main Street in Central Washington's Hispanic places. Spanish language signage and murals that depict Mexican themes adorn the storefronts of Hispanic-owned establishments. Stores and eateries carry products and create dishes that provide resident Hispanics a "taste of home." Store shelves are stocked with herbs, spices, and *masa* packaged in Mexico. Hispanic-owned, and some chain, groceries offer *tortillerias* that produce fresh tortillas daily. These stores also commonly offer a collection of Mexican videocassettes for rent and sell and black baseball caps with names of Mexican states or soccer teams.

Diners and "taco wagons" are the most common Hispanic-owned eateries in Central Washington's Hispanic places. The Main Street diners and bakeries are now *restaurantes* and *panaderias* that serve Mexican entrees and pastries. Taco wagons occupy parking lots and vacant lots, and offer outdoor dining on picnic tables shaded with blue tarpaulins or parasols. Menus include common Mexican dishes such as cow's tongue and brain.

Main Street's storefronts advertise several ways that residents can maintain connections with Mexico. Colorful telephone calling card posters tout special, low rates to Mexico. Spanish language Western Union signs alert workers that they may *"enviar su moneda"* south of the border. Passenger bus company placards publicize schedules of daily departures for Tijuana. Currently, five bus companies, with names such as *Transportes Fronteras* or *Linea Express*, serve the region.

Central Washington's Hispanics are further connected to Mexico and to each other by a handful of weekly Spanish-language newspapers that circulate within the region. Typically, these papers' front pages carry wire stories from Mexico and other Latin American countries. The sports pages allow readers to keep up with Latin American and local soccer leagues. These papers also carry local stories and announcements about Hispanic life in Central Washington.

Spanish-language newspapers also alert readers to leisure activities in the area that are organized specifically for Hispanics, especially soccer and live music. Central Washington's Hispanic places support several soccer leagues. Games typically are played during weekends on school fields. Teams sport uniforms, often emblazoned with names and insignias of Mexican professional teams. Similar to the antecedents of professional sports leagues in the United States, Central Washington's Warehouse Soccer League is comprised of teams sponsored by local fruit companies—on some Saturday, locals might attend the match between the Allan Brothers' *Diablos Rojos* and the team from Yakima's Inland Joseph Fruit Company.

Central Washington's Hispanic places support a live music scene. The region's dance halls have become regular stops on the tours of music groups like *Los Tigres del Norte*. Locals can tune into seven Spanish-language radio stations that play *Norteño* music. On Friday nights, Sunnyside's *Radio Cadeña* broadcasts two hours of *corridas*—songs that recount heroic border crossings among other things. Several schools in the region have contributed to the mix of live music by sponsoring youth *marimba* and *mariachi* bands.

Summary

A visit to one of Central Washington's Hispanic places offers an observer a view of how a population of seasonal and permanent migrants has created a place that feels like home. The experience is intense. Although surrounded by the large Anglo-owned farms on which they earn their wages, Hispanics can live in these settlements without speaking English. They can worship in Spanish. They can eat their customary foods. They can compete in their national sport against their countrymen. They can listen and dance to the music of their homeland.

Upon leaving one of these Hispanic places, traces of migrant ethnicity vanish abruptly. An observer's attention is drawn to a landscape dominated by irrigated agriculture, where the growing Hispanic presence might be overlooked. For miles and miles, large center pivot sprinklers water the formerly shrub-steppe plain and produce expansive hay, alfalfa, hops, and potato fields, apple orchards, and vineyards. Hispanic laborers are an important presence in this massive agricultural landscape and the landlocked islands of intensive Hispanic settlement that they have created are becoming more visible.

Endnotes

1. For a vegetation map of the region, see Meinig 1968, page 172.

2. The 1960 census was the first year that Hispanics were counted. Prior to 1960, information about ethnic minority populations, other than African-Americans, comes from tallies of foreign language speakers.

References

Arreola, Daniel D. and James Curtis. 1993. *The Mexican Border Cities: Landscape Anatomy and Place Personality.* Tucson and London: University of Arizona Press.

Chevalier, François. 1970. *Land and Society in Colonial Mexico: The Great Hacienda.* Berkeley: University of California Press.

Gamboa, Erasmo. 1990. *Mexican Labor and World War II. Braceros in the Pacific Northwest, 1942-1947.* Austin: University of Texas Press.

Gamio, Manuel. 1971. *The Life Story of the Mexican Immigrant.* New York: Dover Publications.

Haverluk, Terrence W. 1998. Hispanic Community Types and Assimilation in Mex-America. *Professional Geographer* 50(4): 465-480.

Maldonado, Carlos S. 1998. An Overview of the Mexicano/Chicano Presence in the Pacific Northwest. In *The Chicana Experience in the Northwest*, eds. Carlos S. Maldonado and Gilberto Garcia. Dubuque: Kendall/Hunt Publishing Company.

Meinig, Donald, W. 1968. *The Great Columbia Plain.* Seattle and London: University of Washington Press.

Therrien, Melissa and Roberta R. Ramirez. 2000. *The Hispanic Population in the United States: March 2000, Current Population Reports,* P20-535, U.S. Census Bureau, Washington, D.C.

United States Department of Commerce, Bureau of the Census. 1930 through 2000 Decennial Censuses. Washington, D.C.

United States Department of Commerce, Bureau of the Census. 1993. *We The Hispanics.* Washington, D.C.

Watt, James W. 1978. *Journal of Mule Train Packing in Eastern Washington in the 1860's.* Fairfield, WA: Ye Galleon Press.

West, Robert C. and John P. Augelli. 1989. *Middle America: Its Lands and Peoples.* Englewood Cliffs: Prentice Hall.

White, Richard. 1995. *The Organic Machine.* New York: Hill and Wang.

Drying up

The Chinese must act fast to conserve their country's shrinking water supply

AS A deputy prime minister in 1999, Wen Jiabao warned that the very "survival of the Chinese nation" was threatened by looming water shortages. Mr Wen has since taken over as prime minister, and earned robust applause in parliament this spring when he promised "clean water for the people". To that end, his government says it will spend an extra $240m this year. But this is a drop in the ocean. Never especially blessed with water, in recent years China has seen its supplies fall to dangerously low levels as it faces drought, rising demand and the combined effects of decades of pollution and misguided policies. Senior officials and international agencies are equally gloomy.

One in three country-dwellers in China lacks access to safe drinking water. More than 100 big cities, of which half are deemed "seriously threatened", are short of water. Water tables are dropping by a metre or more every year across much of northern China. Even in Beijing, supply per head now stands at a perilously low 300 cubic metres (66,000 gallons) a year. Reduced flow rates on China's greatest rivers have made hydro plants reduce badly needed power output: many smelters, paper mills and petrochemical plants are no longer sure of getting the huge amounts of water they require. Droughts, historically more common in northern China, are now hitting the south too. This year Guangdong province, home to 110m people, has had a 40% drop in rainfall.

Misguided pricing policies have made matters a lot worse. Until 1985, most users were not charged at all, so it made little sense for enterprises to invest in treatment and recycling technology or for farmers to fret about wasteful irrigation. Water prices in China have risen only slowly in the past 20 years and are still among the world's lowest. Most Chinese water is bought at around 40% below cost. Even in parched Beijing, city officials are hesitat-

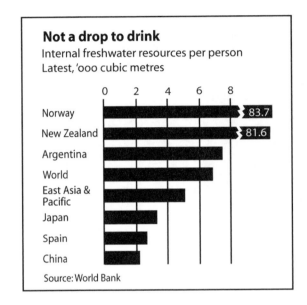

Not a drop to drink
Internal freshwater resources per person
Latest, 'ooo cubic metres

	0	2	4	6	8	
Norway						83.7
New Zealand						81.6
Argentina						
World						
East Asia & Pacific						
Japan						
Spain						
China						

Source: World Bank

ing to press ahead with planned increases to the break-even point of around six yuan ($0.72) a cubic metre. Nationwide, prices are set at different levels for different sorts of users by a jumble of local and central bureaucracies, all wary of slowing economic growth.

John McAlister, head of AquaBioTronic.com, a water-recycling firm, says China is committing "ecological suicide" with its current policies and should put prices up to 20–40 yuan a cubic metre. He argues that foreigners ought to take the lead in arguing for such increases if they hope to continue to benefit from China's economic boom. Alas, he has had as much trouble convincing foreign firms of the urgency of China's water crisis as he has had convincing Chinese firms to invest in his technology.

Living with the
DESERT

Can we learn to love this landscape without killing it?
Here's how one Arizona community found a way

Terry McCarthy, Tucson

A BOBCAT REGULARLY SAUNTERS UP the arroyo leading to Paul and Carolyn Zeiger's desert property in Pima County, Ariz., and leaps onto the flat roof of their adobe-style house. As long as their pet terrier, Stella, is inside, they don't worry much. "The bobcat jumps around up there and takes care of the mice," says Carolyn, 61, a clinical psychologist from Boulder, Colo. The Zeigers also get the occasional rattlesnake on their porch, and in the summer they have to stay indoors to avoid the midday heat. But despite those inconveniences—and in part because of them—they have developed a deep love of the desert in the five years since they moved here. Twenty miles from Tucson, their house looks out on a plain of saguaro cacti stretching to the Rincon Mountains. At night the stars shine brightly without competition from human lighting. Paul, 68, a semiretired software developer, gets all the hiking and bike riding he wants, and Carolyn attends lectures to learn how to grow desert plants in their yard. "You have to learn to adjust in the desert," says Carolyn.

It's an adjustment a lot of folks have been making of late. Since 1950, the population of Arizona, New Mexico and Nevada has increased from 1.6 million to 10 million as Americans discover the desert's clean air, warm weather, open spaces and relatively affordable housing. But without zoning codes to restrict it,

much of that growth has been distressingly haphazard. By the time the Zeigers began looking for a retirement home in the 1990s, what they found was a lot of strip malls, golf clubs and sprawling subdivisions decorated here and there with cactus plants. They were horrified. "We didn't want to move to a place where they are just screwing up the desert again," says Carolyn.

One exception, they discovered, was Pima County, which covers 9,186 sq. mi. of southern Arizona, including the city of Tucson. Pima has developed a conservation plan that permits growth while protecting the desert environment—a plan that has become a template for communities across the Southwest. "The old debate about whether growth is good or bad is irrelevant," says Chuck Huckelberry, Pima County administrator. "We have been growing for 50 years [in Tucson]. But we control where our growth occurs so it maximizes benefits and minimizes impacts."

There's a lot of growth to control. From 1990 to 2003, Arizona's population increased 53%, making it the second fastest growing state in the nation, after Nevada (another desert state, whose population grew 87% in the same period). Developers working in the U.S.'s four major deserts—California's Mojave, Arizona's Sonoran, Texas and New Mexico's Chihuahuan and Nevada and Utah's

Great Basin—can't build houses fast enough. In the town of La Quinta, Calif., southeast of Palm Springs, property prices jumped 48% last year, and new-home buyers have to go on waiting lists or hope to win a developer's lottery for the right to buy a small patch of desert.

> **"If we don't understand how the desert environment works, we will lose it."**
>
> **—William Presch, desert-studies professor**

What those developers are only starting to realize is that deserts are not what they appear to be. Arid, sparsely vegetated and seemingly inhospitable, they look like nature's waste lots, ripe for occupation and improvement. Even the word desert implies "unoccupied." But despite the shortage of water and wide temperature fluctuations, deserts are the host of a wide variety of species, each of which has adapted in its way to life in a desert ecosystem. Couch's spadefoot toads can live underground for much of their lives, awaiting some moisture before they come up and breed. Saguaro cacti are able to suck up a ton of water from one rain shower and then do without more rain for a year. Sidewinder rattlesnakes move across dunes in a unique S-shaped motion that minimizes contact with the scorching sand.

But living on the extremes of viability makes desert creatures surprisingly sensitive to disturbance. Golf courses and suburban lawns soak up sparse groundwater, and indigenous species suffer from the earthmovers, off-road vehicles and domestic pets that new arrivals bring. In California's Coachella Valley in the Mojave Desert, each of the 110 golf courses uses some 750,000 gallons of water a day. "Deserts have fragile ecosystems, and they are being threatened by this development," says William Presch, director of the desert-studies program at California State University at Fullerton. "If we don't understand how the desert environment works, we will lose it."

Once a desert landscape has been despoiled, it recovers slowly, if at all. "Where you have rain, things grow back in 20 or 30 years. In the desert Southwest, it takes centuries," says Huckelberry, an engineer by training. Pima County, he is quick to point out, is not antigrowth—far from it. Every year for the past decade, the population has grown by 15,000 souls and covered 4,500 acres of desert in new housing. The newspapers are flush with property advertisements, the roads out of town dotted with signs for new developments with names like Coyote Creek and Saguaro Buttes. The median single-family-home price in 2004 was $176,500, up 14% from the year before. County planners estimate that the population, now at 943,795, will top 1.6 million by 2050.

But in 1997 the county suddenly found itself paralyzed by a bird, the cactus ferruginous pygmy owl, which was listed as an endangered species after a survey found just 12 of them left in the state. The owl, which weighs 2.5 oz. and nests in cavities in saguaro cacti, had established a small population in prime development land northwest of Tucson. After the bird's listing, house building in the area came to a halt.

Huckelberry decided to use the owl's plight as the impetus to craft a comprehensive conservation plan. He assembled an unlikely coalition of developers, Realtors, ranchers and environmentalists and drew up a blueprint that would protect not just the pygmy owl but a total of

HOT ENOUGH FOR YA?

The desert isn't inhospitable to everyone. These animals thrive in it

Cactus Ferruginous Pygmy Owl

The tiny bird, which nests in cavities in saguaro cacti, is a ferocious hunter, eating much bigger birds if necessary. But its habitat is being destroyed by development. It's listed as endangered

Sidewinder Rattlesnake

Making its home in the Sonoran Desert, the snake takes the 120° heat by moving across the dunes in a unique S-shaped motion that minimizes contact with the hot sand

Couch's Spadefoot Toad

The toad spends much of its life below ground, awaiting rain, then coming up to mate and spawn in the puddles that form. The eggs must hatch and tadpoles turn to toads before the water disappears

55 threatened species—while leaving room for housing development in non-sensitive desert regions. "We believed it was better to be at one table rather than have a huge fight," says Bill Arnold, one of the county's biggest Realtors. "Everyone was a winner in the end."

The Pima County board of supervisors ratified the plan, dubbed the Sonoran Desert Conservation Plan, and last year the county proposed a $174 million bond issue to buy up open land for conservation. The measure passed easily, with 65% voter approval. Under the plan, the county allows concentrated growth in designated areas while preserving swaths of open space in environmentally sensitive core areas in a large ring around Tucson. That open space preserves the characteristic desert vistas while providing corridors for wildlife to move around the edges of housing areas. Huckelberry aims to preserve 263,880 acres in that way.

In addition to the plan, the county has adopted rules governing the exterior colors of new houses, the amount of light that can be given off at night and the amount of water that can be used for gardening. In the 260-unit Arizona Senior Academy, where the Zeigers chose to

settle, grass lawns and water-thirsty plants like oleander are forbidden, there are no streetlights on the roads, and half the land is preserved as open space for wildlife habitat. The houses are built in clusters of four sharing a single driveway and auto court and are designed to be inconspicuous: all exterior walls must mimic the brown and ocher tones of desert soil.

The Sonoran Desert Conservation Plan was written to conserve the region's biodiversity, but Huckelberry concedes that even the most careful planning cannot forestall all the threats that man poses to the desert. Domestic dogs and cats, for example, can wreak havoc on native species of birds and small mammals. One of the main reasons for the pygmy owl's decline was predation by house cats.

An even more serious threat is posed by buffel grass, an invasive species that was originally imported from Kenya to feed cattle. Adapted to being trampled by elephants and capable of spreading widely with little water, buffel grass has migrated west from the rangelands of Texas, bringing a new threat, fire. To conserve water, most desert species in the Southwest grow far apart, making it hard for fires to spread. Buffel grass grows easily in dry soil, forming a carpet of dry, flammable stalks that burns very hot after a lightning strike and can engulf cacti, yucca, ocotillo and the paloverde trees. "None of the native plants have fire adaptation. If they burn, they die," says Tom Van Devender, a senior research scientist at the Arizona-Sonora Desert Museum in Tucson. "If there is recurring fire, you get a conversion from desert to savannah grassland."

But the most fundamental limitation to life in the desert is water, and no matter how sensitively houses are located aboveground, everyone is still drawing from the same precious supply of groundwater. In the long term, overuse of groundwater means slow death for desert plants, whose roots are unable to reach down far enough to sustain them. When plants die, animals run out of food and shelter—a process that is often noticed only after it's too late.

The idea of treating the desert gently, pioneered by forward-thinking planners like Huckelberry, is finally starting to catch on. "The conservation issue has just exploded," says Mike Chedester, education curator for the Living Desert University in Palm Desert, Calif. The program began only three years ago, and now he runs 124 courses a year on desert ecology and xericulture, or gardening with desert plants. "We have many students who come out to the desert, buy a home prelandscaped with lawns that need watering two or three times a day,

and after a while they realize it doesn't make sense."

Carolyn Zeiger is doing her best to reduce the impact their home makes on the desert. "Given the rate the desert is being gobbled up by people like us, my feeling is we need to put some back," she says, standing on her porch and pointing to the plants in her yard. "I put in native plants only—ocotillo, Arizona rosewood, desert willow, prickly pear. I start them

with a little water, but soon they will survive on their own."

The desert stretches out in front of her, the ground turning pink in the sunlight and the distant mountains a dark shade of blue. The desert doesn't need much to survive—a little moisture, not too much disturbance, a little respect from humans. Then it too can survive on its own.

From *Time Magazine*, April 4, 2005, pp. 46-49. Copyright © 2005 by Time Inc. Reprinted by permission.

deep blue thoughts

We need the ocean, and now more than ever, the imperiled ocean needs us

Sylvia A. Earle

"Who needs the ocean?" a reporter in Australia once asked me, then added, "I don't eat fish. I don't swim, I get sea sick. People don't drink salt water. Why should I care whether the ocean exists or not?

Here are a few good reasons: The ocean generates more than 70 percent of the oxygen in the atmosphere, absorbs much of the carbon dioxide, shapes planetary chemistry, stabilizes temperature, drives climate and weather, replenishes fresh water to land and sea through clouds that form and return water as rain, sleet and snow. It is home for most of life on earth, comprising 97 percent of the biosphere—a three-dimensional realm that has an averaged depth of two-and-a-half miles, and a maximum depth of seven miles. Every spoonful is filled with living creatures, some large enough to see, but mostly microscopic. I think of those creatures as the basic ingredients in a living minestrone that make that world work as it does.

Quite simply, water is the key to life, the cornerstone of earth's life support system, the single non-negotiable thing life requires. We tend to prize fresh water, a substance presently acknowledged to be more valuable even than oil—and certainly more costly, ounce for ounce, than oil when sold by the bottle, even at 2004 prices. The ocean is mostly salt water, but it is not *just* salt water. Our bodies are mostly salt water, but not *just* salt water. The sea functions as it does, generates oxygen as it does, absorbs carbon dioxide as it does, because it is basically filled from top to bottom with countless small, medium and sometimes very large creatures that are continuously processing sunlight, water, minerals and chemicals in the water and one another in a huge eat-and-be-eaten arena that is mostly cold—near freezing below 1,000 feet in a medium that has an average depth of 14,000 feet, a maximum of seven miles.

Below a few hundred feet, the sea is eternally, infinitely dark, the deep, utter blackness of dreamless sleep punctuated by sudden flashes of eerie, blue-green light. Ninety percent of the creatures in the deep sea sparkle or glow with their own, living luminesence, sometimes creating soft, slow pulses, sometimes exploding with startling brilliance reminiscent of the 4th of July.

IN LESS THAN 50 YEARS, MANY COMMERCIALLY EXPLOITED SPECIES HAVE BECOME SCARCE, DESPITE INCREASED EFFORT, NEW METHODS TO CAPTURE DEEPER DWELLING POPULATIONS, AND A CONSTANT SHIFTING OF TARGETS.

For me, diving into the sea is like diving into the history of life on earth. Barely beneath the surface, I am immersed in a bouillabaisse of diaphanous jellies, tiny, translucent fish flashing glints of silver and thousands of nameless, glistening specks, the young of many of the major divisions of animal life on earth—tiny, jewel-like squids; miniature rippling, whirling worms dancing to their own silent rhythms; pea-sized baby crabs that would terrify Godzilla if they were Godzilla-sized. These are creatures whose history precedes that of dinosaurs by more than a hundred million years, and there they are, descendents of hardy ancestors that miraculously survived times when the planet was much colder than it is today, times when it was much warmer, and even the time when the Earth reeled from the impact of an asteroid or whatever it was that did in the dinosaurs. As a child, I thought of the ocean as endlessly wondrous, the wildest of wild places, all-encompassing, infinitely mysterious, ever changing—yet changeless, in the sense that there wasn't anything people could do that could alter the way it was. You could put whatever you wanted into the sea, and it would simply disappear, forever. At the same time, you could take out as many fish or crabs or dams as you cared to. How could the ocean ever run out of fish? It did not occur to me that I would personally witness a "sea change" in the nature of the ocean itself, from a Sea of Eden to something perilously close to Paradise Lost. I hadn't figured on human population tripling in my lifetime, and with it, a more than tripling in demand for seafood and industrial-scale extraction of fish meal and oil, nor the capacity of my species to put enough wastes and incidental flow of noxious materials into the ocean to alter its basic chemistry.

Over the ages, people have tended to regard the sea, like the skies above, as "free." Like the air, the sea is part of what some call the global commons, immense blue spaces that both separate the people of the world—and join them. Traditions have grown, notably the concept of "Freedom of the Seas"—which holds that the resources of the high seas beyond coastal regions claimed by nations must be accessible to all states and their fleets on equal terms, Until recently, little thought was given to the amount of fish or other wildlife that could or should be extracted, whether near shore or in the distant, open ocean because of the seductive notion that, miraculously, there would always be more.

However, the reality is quite different. In less than 50 years, many commercially exploited species have become scarce, despite increased effort, new methods to capture deeper dwelling populations, and a constant shifting of targets. Fish once spurned such as pollock, monkfish, sharks, and capelin have become prime prey. Orange roughy taken from cold, dark depths of 2,000 feet and more near New Zealand take 30 years to mature and may be a century old by the time they are caught and sent to a supermarket or restaurant. Forty years ago they were protected by default because of their inaccessibility, and because the means to find them and ship them worldwide did not exist, but no more. After years of being commercially consumed, they are in sharp decline, along with other deep-sea fish such as Chilean sea bass and grenadiers, commonly sold as "hoki." Sharks—feared by many as the ultimate predators in the sea-have more than met their match in us. For every shark that nibbles on a human (several times a year) millions of sharks are munched on by humans. Cod, economic mainstay of several Atlantic nations for 500 years, have been brought to the edge of annihilation in the past quarter century.

HUNTERS IN THE SEA HAVE LESS FINESSE THAN THEIR TERRESTRIAL COUNTERPARTS, USING TRAWLS AND NETS TO CAPTURE MOSTLY CARNIVOROUS FISH, SHRIMP, SCALLOPS AND OTHER WILDLIFE—THE EQUIVALENT OF USING BULLDOZERS TO CATCH SQUIRRELS.

Part of the problem relates to the methods used to catch marine life. Hunters in the sea have less finesse than their terrestrial counterparts, using trawls and nets to capture mostly carnivorous fish, shrimp, scallops and other wildlife—the equivalent of using bulldozers to catch squirrels. Such techniques result in a lot of collateral damage in terms of habitat destroyed and non-targeted creatures unintentionally trapped, killed and discarded, often with a ratio of ten to one: one pound of intended catch, ten pounds of "by-catch." Consuming large, old carnivores involves another kind of by-catch, that is, the tremendous hidden food-chain cost needed to make such things as tuna, sharks and lobsters; thousands of pounds

of plants at the bottom of a long and complex web of life are invested in every pound of a 10-year-old grouper, swordfish or tuna. Few could afford to eat many of the wild species now widely marketed if the true ecosystem cost were accounted for.

When I served as chief scientist for the National Oceanic and Atmospheric Administration (which includes the National Marine Fisheries Service) in the early 1990s, I was shocked to discover that 90 percent of the blue fin tuna in the North Atlantic had been taken by commercial fishing. Now, according to the recent analysis of decades of data, it appears that fishermen are so good at finding and catching ocean wildlife that in half a century, 90 percent of the large, most sought-after fish have been eliminated—several kinds of tuna, swordfish, marlin, sharks, cod, grouper, snapper, halibut, and numerous others. Most of the ocean's big fish are gone.

MUCH OF THE OCEAN REMAINS IN GOOD HEALTH AND CAN HELP STABILIZE AND RESTORE DAMAGED AREAS AND PERHAPS FORESTALL CASTASTROPHIC CHANGES—IF ACTIONS ARE TAKEN SOON.

The good news is that 10 percent remain. While about half of the world's coral reefs have disappeared or seriously declined in the past 50 years, half are still in pretty good shape. More than 150 "dead zones" have developed in coastal waters, globally, a consequence of upstream sources of excess fertilizers and other unwelcome additions to the sea. Accelerated global warming looms large, an overarching source of trouble for ecosystems already badly stressed, with the potential to alter ocean currents, climate, weather, and much that matters directly to human beings. However, much of the ocean remains in good health and can help stabilize and restore damaged areas and perhaps forestall catastrophic changes—if actions are taken soon.

The United Nations Law of the Sea Convention, now in effect and endorsed by most nations, has some provisions for the protection of marine resources beyond areas of national jurisdiction, but does not adequately address many current problems, and lacks the means for enforcement. It also lacks ratification by the United States. However, the US did recently undertake the first comprehensive assessment of national ocean policy in more than three decades, first as a privately-funded initiative conducted by the Pew Oceans Commission, and soon thereafter, by the official US Commission on Ocean Policy, appointed by the president in 2001. The US Commission's report, released in 2004, criticized our past approach to the oceans and urged immediate steps to formulate a coherent, comprehensive and effective national ocean policy.

The need for a global strategy and practical plan of action inspired an international meeting of scientists, business and industry leaders, educators, media experts, economists and government officials from 20 countries and more than 70 organizations to meet in Los Cabos, Mexico in 2003, after a year of preliminary research and preparation. They were charged with the goal of developing a plan of action—a serious business plan, with timelines, milestones and costs to define what could be done to halt the troublesome trends. Called "Defying Ocean's End," the conference proposed the following ideas for a 10-year plan that can shift the present cataclysmic slide: reforming fisheries to develop sustainable practices; expanding research on top-priority marine environments and pollution sources; creating and strengthening marine protected areas; implementing local, regional and global communication and education programs to engage the public and promote awareness of why the ocean matters; and enacting more active governance of the oceans.

The atmosphere at the conference was charged with optimism; the meetings sizzled with the awareness that real solutions were possible. The plan, when fully developed, could make a real difference. But what will it cost?

A team of business professionals, working with the other experts assembled at Los Cabos, concluded that significant progress could be made in terms of "defying ocean's end" by mobilizing private and public resources, about two billion dollars a year, for 10 years. Some observed that redirecting a portion of the $50 billion or so spent on fisheries subsidies every year could help both the fish and the fishermen. Others focused on how much more it will cost if we do nothing.

Whatever happens, there is little doubt that this is a time of magnified significance for the future of the ocean—and for ourselves. Although some, like the "I-don't-swim, don't-drink-salt water" Australian reporter, do not yet see why the ocean matters, many have come to understand the magnitude of the problems, and the consequences of complacency. In the next 10 years, as never before, we have a chance to get it right—and for coral reefs, for giant tuna and the ocean I knew as a child, maybe as never again.

From *The Aspen Idea,* Winter 2004/2005, pp. 54-57. Copyright © 2005 by Sylvia A. Earle. Reprinted by permission of the author.

An Inner-City Renaissance

The nation's ghettos are making surprising strides. Will the gains last?

Take a stroll around Harlem these days, and you'll find plenty of the broken windows and rundown buildings that typify America's ghettos. But you'll also see a neighborhood blooming with signs of economic vitality. New restaurants have opened on the main drag, 125th Street, not far from a huge Pathmark supermarket, one of the first chains to offer an alternative to overpriced bodegas when it moved in four years ago. There's a Starbucks—and nearby, Harlem U.S.A., a swank complex that opened in 2001 with a nine-screen Magic Johnson Theatres, plus Disney and Old Navy stores and other retail outlets. Despite the aftermath of September 11 and a sluggish economy, condos are still going up and brownstones are being renovated as the middle classes—mostly minorities but also whites—snap up houses that are cheap by Manhattan standards.

It's not just Harlem, either. Across the U.S., an astonishing economic trend got under way in the 1990s. After half a century of relentless decline, many of America's blighted inner cities have begun to improve. On a wide range of economic measures, ghettos and their surrounding neighborhoods actually outpaced the U.S. as a whole, according to a new study of the 100 largest inner cities by Boston's Initiative for a Competitive Inner City, a group founded in 1994 by Harvard University management professor Michael E. Porter.

Consider this: Median inner-city household incomes grew by 20% between 1990 and 2000, to a surprising $35,000 a year, the ICIC found, while the national median gained only 14%, to about $57,000. Inner-city poverty fell faster than poverty did in the U.S. as a whole, housing units and homeownership grew more quickly, and even the share of the

population with high school degrees increased more. Employment growth didn't outdo the national average, with jobs climbing 1% a year between 1995 and 2001, vs. 2% nationally. Still, the fact that inner cities, which are 82% minority, created any jobs at all after decades of steady shrinkage is something of a miracle.

SCENT OF OPPORTUNITY

NOR ARE THE GAINS just the byproduct of the superheated economy of the late 1990s. Rather, they represent a fundamental shift in the economics of the inner city as falling crime rates and crowded suburbs lure the middle-class back to America's downtowns. After decades of flight out of inner cities, companies as diverse as Bank of America, Merrill Lynch, and Home Depot have begun to see them as juicy investment opportunities. National chains are opening stores, auto dealerships, and banks to tap into the unfulfilled demand of inner cities.

Wall Street, too, is jumping in, making loans and putting up equity for local entrepreneurs. "Smart businesspeople gravitate toward good opportunities, and it has become clear that inner cities are just that," says David W. Tralka, chairman of Merrill Lynch & Co.'s Business Financial Services group. In 2002, his group, which caters to small business, began formally targeting inner cities. It now offers financing and commercial mortgages for hundreds of inner-city entrepreneurs around the country.

Is it possible that America at last has started to solve one of its most intractable social ills? True, the progress so far is minuscule compared with the problems created by decades of capital flight, abysmal schools, and drug abuse. And some inner cities, like Detroit's, have made little sus-

tained progress. Ghettos also have been hit by the joblessness of this latest recovery. The national poverty rate has jumped by nearly a percentage point since 2000, to 12.1% last year, so it almost certainly did likewise in inner cities, which the ICIC defined as census tracts with poverty rates of 20% or more.

But as the economy recovers, a confluence of long-term trends is likely to continue to lift inner cities for years. The falling crime rate across the country has been a key factor, easing fears that you take your life in your hands by setting foot in an inner city. At the same time, larger demographic shifts—aging boomers turned empty nesters, more gays and nontraditional households without children, homeowners fed up with long commutes—have propelled Americans back into cities. When they arrive, slums suddenly look like choice real estate at bargain prices.

BEYOND PHILANTHROPY

POLITICAL AND CIVIC LEADERS helped lay the groundwork, too. After floundering for decades following the exodus of factories to the suburbs in the 1950s, many cities finally found new economic missions in the 1990s, such as tourism, entertainment, finance, and services. This has helped boost the geographic desirability of inner-city areas. New state and federal policies brought private capital back, too, by putting teeth into anti-redlining laws and by switching housing subsidies from public projects to tax breaks for builders. As a result, neighborhoods like the predominantly African-American Leimert Park in South Central Los Angeles are becoming thriving enclaves.

The outcome has been a burst of corporate and entrepreneurial activity that already has done more to transform inner

Inner Cities and Their Residents ...

The Boston-based Initiative for a Competitive Inner City has completed the first-ever analysis of the 100-largest inner cities in the U.S. and finds the once-dismal picture brightening

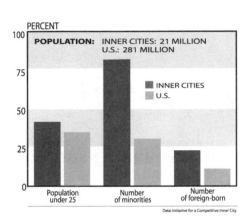

PERCENT

POPULATION: INNER CITIES: 21 MILLION
U.S.: 281 MILLION

■ INNER CITIES
▫ U.S.

Population under 25 | Number of minorities | Number of foreign-born

Data: Initiative for a Competitive Inner City

...Did Better than the Nation As a Whole in the 90s...

Change between 1990 and 2000 Census data

	INNER CITIES	U.S.
Population	24%	13%
Household income	20%	14%
Housing unit growth	20%	13%
High school graduates*	55 to 61%	75 to 80%
College graduates*	10 to 13%	20 to 24%
Home ownership	29 to 32%	64 to 66%
Poverty rate	34 to 30%	13 to 12%

*Of those 25 and over

...Although There's Still A Long Way to Go

2000 Census data

	INNER CITIES	U.S.
High school graduates*	61%	80%
College graduates*	13%	24%
Poverty rate	31%	11.3%
Unemployment rate	12.8%	5.8%
Home ownership rate	32%	66%
Average household income	$34,755	$56,600
Aggregate household income	$250 billion	$6 trillion

Data: Initiative for a Competitive Inner City

cities than have decades of philanthropy and government programs. "What we couldn't get people to do on a social basis they're willing to do on an economic basis," says Albert B. Ratner, co-chairman of Forest City Enterprises Inc., a $5 billion real estate investment company that has invested in dozens of inner-city projects across the country.

EMERGING MARKETS

THE NEW VIEW OF GHETTOS began to take hold in the mid-1990s, when people such as Bill Clinton and Jesse Jackson started likening them to emerging markets overseas. Porter set up the ICIC in 1994 as an advocacy group to promote inner cities as overlooked investment opportunities. Since then, it has worked with a range of companies, including BofA, Merrill Lynch, Boston Consulting Group, and PricewaterhouseCoopers to analyze just how much spending power exists in inner cities.

The new study, due to be released on Oct. 16, uses detailed census tract data to paint the first comprehensive economic and demographic portrait of the 21 million people who live in the 100 largest inner cities. The goal, says Porter, "is to get market forces to bring inner cities up to surrounding levels."

Taken together, the data show an extraordinary renaissance under way in places long ago written off as lost causes. America's ghettos first began to form early in the last century, as blacks left Southern farms for factory jobs in Northern cities. By World War II, most major cities had areas that were up to 80% black, according to the 1993 book *American Apartheid*, co-authored by University of Pennsylvania sociology professor Douglas S. Massey and Nancy A. Denton, a sociology professor at the State University of New York at Albany. Ghettos grew faster after World War II as most blacks and Hispanics who could follow manufacturing jobs to the suburbs did so, leaving behind the poorest and most

un-employable. Immigrants poured in, too, although most tended to leave as they assimilated.

In this context, the solid gains the ICIC found in the 1990s represent an extraordinary shift in fortunes. One of the biggest changes has come in housing. As cities have become desirable places to live again, the number of inner-city housing units jumped by 20% in the 1990s, vs. 13% average for the U.S. as a whole.

A number of companies were quick to see the change. BofA, for example, has developed a thriving inner-city business since it first began to see ghettos as a growth market six years ago. In 1999 it pulled together a new unit called Community Development Banking, which focuses primarily on affordable housing for urban, mostly inner-city, markets, says CDB President Douglas B. Woodruff. His group's 300 associates are on track this year to make $1.5 billion in housing loans in 38 cities, from Baltimore to St. Louis.

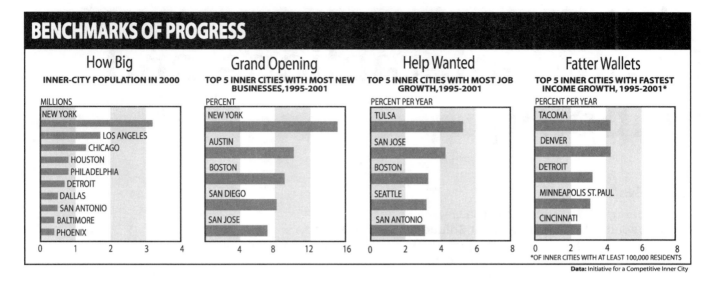

BENCHMARKS OF PROGRESS

How Big
INNER-CITY POPULATION IN 2000

Grand Opening
TOP 5 INNER CITIES WITH MOST NEW BUSINESSES, 1995-2001

Help Wanted
TOP 5 INNER CITIES WITH MOST JOB GROWTH, 1995-2001

Fatter Wallets
TOP 5 INNER CITIES WITH FASTEST INCOME GROWTH, 1995-2001*

*OF INNER CITIES WITH AT LEAST 100,000 RESIDENTS

Data: Initiative for a Competitive Inner City

They will do an additional $550 million in equity investments, mostly real estate.

SHELLED OUT

PENSION FUNDS AND other large investors are putting in cash, too. The Los Angeles County Employee Retirement Assn. has sunk $210 million into urban real estate since 2000, including $87 million in August for a bankrupt, 2,496-room apartment complex in Brooklyn, N.Y. The plan is to do things like fix the broken elevators, hire security guards, and kick out nonpaying tenants. "We believe there are opportunities that weren't there before or that we weren't aware of," says board member Bruce Perelman.

One question is whether the ICIC's findings represent not so much progress by the poor as their displacement by middle-class newcomers. In other words, inner-city incomes could be rising simply because affluent new home buyers jack up the average. But experts think gentrification explains only a small part of what's going on. "It's certainly a local phenomenon, but if you aggregate 100 inner cities, gentrification is a small trend," says ICIC research director Alvaro Lima, who spearheaded the study.

In Chicago, for instance, a $65 million redevelopment of the notorious Cabrini-Green housing project has replaced three slummy high-rises with mixed-income units. The area has a new library, new schools, and a new retail center featuring a major grocery store, Starbucks, and Blockbuster (BBI)—all staffed by scores of local residents. "The goal is not gentrification, it's to integrate the classes," says Phyllis L. Martin, the head of a local committee that's trying to lure more than $50 million in private capital to help the city replace 3,245 public housing units in another blighted area, Bronzeville.

"We're just beginning to undo all the damage"

Despite the brightening picture, the decay of most inner cities is so advanced that half a dozen years of progress makes only a dent. The degree of poverty—a measure of how many poor people there are in a census tract—fell 11% in 60 large cities in the 1990s, according to an analysis by U Penn's Massey that parallels ICIC's approach. While that's a significant decline, it only begins to offset the doubling of poverty concentrations in prior decades, he found. "The gains are the first positive news since at least the 1950s, but we're just beginning to undo all the damage," says Massey.

BADGE OF SHAME

WHAT'S MORE, too many inner cities remain untouched. More than a third of the ICIC's 100 cities lost jobs between 1995 and 2001. Detroit's ghetto has seen little new development and shed one-fifth of its jobs over this period. Residents did gain from the booming auto industry, which hired many locals and pushed up their median incomes at a 3.2% annual pace in the 1990s—the third highest increase of the ICIC 100. But with auto makers now shedding jobs again, those gains are likely to be short-lived. More broadly, improving inner cities won't come close to wiping out poverty in the U.S. While the inner-city poverty rate of 31% is nearly three times the national average, the 6.5 million poor people who live there represent less than a fifth of the country's 34.6 million poor.

Still, America's ghettos have been a national badge of shame for so long that any real gain is news. The change in perspective also seems to be an enduring one, not just a 1990s blip. For evidence, consider Potamkin Auto Group, which owns 70 dealerships around the country and will break ground in Harlem in late October on a $50 million development that will include Cadillac, Chevrolet, Hummer, and Saturn dealers. Potamkin also has a project in another inner city and is mulling a national expansion. "We see opportunities there," says Robert Potamkin, president of the family-owned company. This view, that inner cities can be a good place to do business, may be the most hopeful news about the country's urban blight in decades.

—By Aaron Bernstein in Washington, with Christopher Palmeri in Los Angeles and Roger O. Crockett in Chicago

UNIT 4

Spatial Interactions and Mapping

Unit Selections

Key Points to Consider

- Describe the spatial form of the place in which you live. Do you live in a rural area, a town, or a city, and why was that particular location chosen?

- How does your hometown interact with its surrounding region? With other places in the state? With other states? With other places in the world?

- Discuss the importance of the geographical concept of accessibility.

- What problems occur when transportation systems are overloaded?

- How will public transportation be different in the future? Will there be more or fewer private autos in the next 25 years? Defend your answer.

- In what ways has telecommunication and the Internet brought world regions "closer" to each other?

- How can a balance be achieved between economic development and environmental conservation?

- How good a map-reader are you? Why are maps useful in studying a place?

Student Website
www.mhcls.com/online

Internet References
Further information regarding these websites may be found in this book's preface or online.

Edinburgh Geographical Information Systems
http://www.geo.ed.ac.uk/home/gishome.html

Geography for GIS
http://www.ncgia.ucsb.edu/cctp/units/geog_for_GIS/GC_index.html

GIS Frequently Asked Questions and General Information
http://www.census.gov/geo/www/faq-index.html

International Map Trade Association
http://www.maptrade.org

PSC Publications
http://www.psc.isr.umich.edu

Geography is the study—not only of places in their own right but also of the ways in which places interact. Highways, airline routes, telecommunication systems, and even thoughts connect places. These forms of spatial interaction are an important part of the work of geographers.

Three articles discuss the importance of GIS within the context of geographical concepts. "Mapping Opportunities," discusses the importance of GIS within the context of geographical concepts. The next asks if publicly available geospatial information presents a homeland security risk in the age of terrorism. The article "Calling all Nations" uses varying line width on a map to report the degree of communication between the U.S. and other countries. Next, "Mapping the Diversity of Nature" reviews an important research project in Middle America. Naval maps produced during World War II are reviewed in the next selection.

"Asphalt and the Jungle" proposes that transportation development in Brazil's rainforest can bring economic growth without endangering the fragile environment. Managua, Nicaragua, devastated by an enormous earthquake in 1972, is a city literally without a map. The last article discusses the continued devastation of AIDS in southern Africa.

It is essential that geographers be able to describe the detailed spatial patterns of the world. Neither photographs nor words could do the job adequately, because they literally capture too much of the detail of a place. Therefore, maps seem to be the best way to present many of the topics analyzed in geography. Maps and geography go hand in hand. Although maps are used in other disciplines, their association with geography is the most highly developed.

A map is a graphic that presents a generalized and scaled-down view of particular occurrences or themes in an area. If a picture is worth a thousand words, then a map is worth a thousand (or more!) pictures. There is simply no better way to "view" a portion of Earth's surface or an associated pattern than with a map.

Mapping opportunities

**Scientists who can combine geographic information systems
with satellite data are in demand in a variety of disciplines.
Virginia Gewin gets her bearings.**

VIRGINIA GEWIN

Forest fires ravaging southern California, foot-and-mouth disease devastating the British livestock industry, the recent outbreak of severe acute respiratory syndrome (SARS)—all of these disasters have at least one thing in common: the role played by geospatial analysts, mining satellite images for information to help authorities make crucial decisions. By combining layers of spatially referenced data called geographic information systems (GIS) with remotely sensed aerial or satellite images, these high-tech geographers have turned computer mapping into a powerful decision-making tool.

Natural-resource managers aren't the only ones to take notice. From military planning to real estate, geospatial technologies have changed the face of geography and broadened job prospects across public and private sectors.

Earlier this year, the US Department of Labor identified geotechnology as one of the three most important emerging and evolving fields, along with nanotechnology and biotechnology. Job opportunities are growing and diversifying as geospatial technologies prove their value in ever more areas.

The demand for geospatial skills is growing worldwide, but the job prospects reflect a country's geography, mapping history and even political agenda. In the United States, the focus on homeland security has been one of many factors driving the job market. Another is its vast, unmapped landscape. While European countries are integrating GIS into government decision-making, their well-charted lands give them little need for expensive satellite imagery.

AN EXPANDING MARKET

All indications are that the US$5-billion worldwide geospatial market will grow to $30 billion by 2005—a dramatic increase that is sure to create new jobs, according to Emily DeRocco, assistant secretary at the US Department of Labor's employment and training division. NASA says that 26% of its most highly trained geotech staff are due to retire in the next decade, and the National Imagery and Mapping Agency is expected to need 7,000 people trained in GIS in the next three years.

Of the 140,000 organizations globally that use GIS, most are government agencies—local, national and international. A ten-year industry forecast put together last year by the American Society for Photogrammetry & Remote Sensing (ASPRS) identified environmental, civil government, defence and security, and transportation as the most active market segments.

Business at the Earth-imagery provider Space Imaging, of Thornton, Colorado, increased by 70% last year, says Gene Colabatistto, executive vice-president of the company's consulting service. To keep up momentum, the company plans to hire more recruits with a combination of technical and business skills. Colabatistto cites the increased adoption of GIS technologies by governments as a reason for the rise. He adds that the US military, the first industry to adopt GIS and remote sensing on a large scale, has spent more than $1 billion on commercial remote sensing and GIS in the past two years.

LOOKING DOWN IS LOOKING UP

The private sector hasn't traditionally offered many jobs for geographers, but location-based services and mapping—or 'geographic management systems'—are changing the field. "The business of looking down is looking up," says Thomas Lillesand, director of the University of Wisconsin's Environmental Remote Sensing Center in Madison, Wisconsin.

Imagery providers such as Digital Globe of Longmont, Colorado, also need more GIS-trained workers as markets continue to emerge. Spokesman Chuck Herring says that the company has identified 54 markets in which spatial data are starting to play a role.

The Environmental Systems Research Institute (ESRI), in Redlands, California, sets the industry standards for geospatial software. Most of its 2,500 employees have undergraduate training in geography or information technology, although PhDs are sought after to fill the software-development positions. Many private companies, including the ESRI and Space Imaging, offer valuable work experience to both graduate and final-year undergraduate students.

Graduates in natural-resource management note that GIS and remote-sensing skills are becoming as important as fieldwork. GIS platforms, which manipulate all forms of image data, are transforming disciplines such as ecology, marine biology and forestry.

"Science has discovered geography," says Doug Richardson, executive director of the Association of American Geographers (AAG). Many of the National Science Foundation's multidisciplinary research programmes now include a geospatial component.

SKILLED LABOUR

Some universities are offering two-year non-thesis master's programmes in geospatial technologies, including communication and business courses—perfect for professionals who want to build on existing skills or move into a new field. The non-profit Sloan Foundation has funded several geospatially related professional master's programmes. In addition, numerous short courses are available to bring professionals up to speed. Indeed, the ESRI alone trains over 200,000 people a year. AAG and ASPRS conferences also offer training sessions.

Although technical skills are important, Richardson stresses that employees need a deep understanding of underlying geographic concepts. "It's a mistake to think that these technologies require only technician-oriented functions," he says.

Throughout the European Union (EU), the many top-quality graduate geography programmes remain the primary training grounds. Recently, a few pan-European projects have also emerged, including a new international

institute designed to train future geographers. Building on a collaboration between the European Space Agency and the US National Science Foundation, the Vespucci Initiative in 2002 began three-week summer workshops training students from around the world in spatial data infrastructure, spatial analysis and geodemographics. The EU even promotes distance learning: UNIGIS, a network of European universities, prides itself on being the only virtual, global, multilingual GIS programme in the world.

Although GIS is increasingly incorporated into UK government practices, there is little demand for remote-sensing expertise in this small and heavily mapped country. Mark Linehan, director of the London-based Association for Geographic Information, says that although the public-sector market is growing, it remains a struggle to find jobs for MScs at the appropriate pay scale and qualification level.

The European Commission (EC) is laying the groundwork to ease data-sharing across countries in anticipation of wider adoption of GIS among the member-state governments and to cut the costs of data gathering. That process alone will require at least a couple of thousand people trained in GIS, and many more proposals are expected.

Indeed, the EC and the European Space Agency have joined to propose a Global Monitoring for Environment and Security initiative, to provide permanent access to information on environmental management, risk surveillance and civil security. Given the scope of the mandate, this is likely to need people who understand how to interpret, integrate and manage satellite information—those who also have a background in natural-resource issues will be in highest demand.

Considering the role that GIS played in staving the spread of foot-and-mouth disease, such a system will not only increase the prevalence of geospatial skills in Europe, it will better connect data with Europe's resource managers.

Virginia Gewin is a freelance science writer in Corvallis, Oregon.

Geospatial Asset Management Solutions

How do we maintain and repair the nation's infrastructure, without going broke?

Damon D. Judd

Typical asset-intensive organizations that have an extensive infrastructure to build, maintain, repair, replace and ultimately decommission, require a substantial amount of work to be managed. In most cases they have or will invest in some type of asset and/or work management system. This is generally a database application that allows facilities, equipment, vehicles, and materials to be tracked and to enable work to be assigned to work crews, managed by their supervisors, and reported to upper management in some fashion.

There are several business drivers that dictate when work needs to be assigned and completed to maintain, repair, and operate the assets owned (or managed) by the organization. Some of those drivers include homeland security concerns, prevention of massive infrastructure failures such as the recent blackout in the northeast U.S., adherence to GASB 34 standards for governmental accounting practices, the effects of deregulation on the utility industry, and compliance with various environmental laws and regulations.

What is GIS Integration with Asset Management?

By integrating the capabilities of a Geographic Information System (GIS), which many organizations already have in place, with an Asset Management System (AMS), public works departments, investor-owned utilities, and other asset-intensive organizations can improve their ability to manage their asset inventories, and do so even more efficiently than ever before. The GIS enables map-based views of the asset and work information that is managed using an AMS.

Because there is a database relationship between the spatial data in the GIS and details of the assets and the work performed on those assets in the AMS, extended graphical display and data analysis functionality becomes possible. In an integrated solution, the graphical perspective can be presented along with the current condition of the selected set of assets. Asset-related information can be spatially analyzed to help identify trends or to determine impacts of proposed operations. Analysis results and trends can be displayed on a map to

further assist in the decision making process.

Where are the Benefits?

- *Executive Decision Support* (by providing current reporting of asset condition levels and map-based views of the infrastructure assets).
- *Customer Service* (by enhancing the customer response level and the details of work orders in a graphical context from the dispatcher to the field worker, such as a call center).
- *Mobile Work Orders* (by putting current maps with asset details and that day's work orders directly in the service truck on a mobile device such as a tablet, ruggedized laptop, or PDA).
- *Capital Projects* Budgeting and planning of capital projects can be improved by supporting the analysis of various alternatives prior to design. Construction management can benefit from having current maps and asset details for the existing and proposed infrastructure in the project's geographic area of interest.
- *Operations and Maintenance Supervisors* can more easily

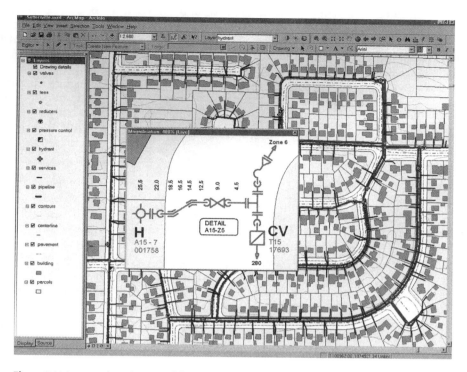

Figure 2 Using map-based views of the infrastructure to support decision making and reporting.

plan and manage field crews by knowing what equipment is needed and where. Maintenance workers can use daily or weekly work assignments that include routes and directions to the job, detailed reports of asset work history and condition levels, and equipment needed for the scheduled work.

- *Finance and Accounting* By tracking what work is completed where, budgets can be compared against actual costs of maintenance and repairs, cost overruns can be avoided through better planning and tracking of work and asset conditions, and continuous improvement of the infrastructure value can be achieved.

Why Bother?

- Cost savings to the organization can be realized by spatially enabling the asset management system.

- Often the investment in implementing a GIS has already been made.
- Extending the integrated system throughout the enterprise, including field services, offers many additional benefits to the organization.

Obstacles to Implementation

Unfortunately, there can be many reasons not to geospatially enable an AMS. Most of those obstacles are based on the total cost to implement, security concerns, lack of data standards (or the availability of metadata), complexity of operation (real or perceived), difficulty in coordination and cooperation between multiple groups, and resistance to change.

Many of these obstacles can be overcome by using good project management principles and by incorporating a change management approach that minimizes the risks of introducing new technology into an organization. The downside of

not implementing such a solution must also be considered as part of the long-range strategic planning process for any organization that is asset-intensive in its operations.

"In the old world that is passing, in the new world that is coming, national efficiency has been and will be a controlling factor in national safety and welfare."
—Gifford Pinchot

Conclusions

By integrating the spatial component using a geographic information system (GIS), the asset knowledge base inherent in an organization's asset management system (AMS) can be better managed, shared, and visualized. Maps with current information become readily available to everyone from field workers to upper management.

Most of the obstacles to implementing such a solution are not difficult technical challenges, but rather require a good change management approach. The changes to the affected business processes that result from a geospatially-enabled AMS will likely lead to continuous process improvement, thereby reducing maintenance costs, improving customer service, enhancing operational efficiency, and increasing the return to the bottom line.

About the Author

Damon D. Judd is President of Ala Carto Consulting, a private consulting practice in Louisville, Colorado offering GIS and spatial data management services to utilities, energy, local government, and environmental organizations.

Internet GIS: Power to the People!

A Web-based GIS provides a public-involvement tool for airport development

By Bernardita Calinao and Candace Brennan

A NEW AIRPORT RUNWAY CAN HAVE A MAJOR EFFECT ON A COMMUNITY. BUT HOW CAN CITIZENS BECOME MORE KNOWLEDGEABLE ABOUT SUCH ISSUES, AND HOW CAN THEY LET THEIR VOICES BE HEARD? IN ERIE, PA., C&S ENGINEERS USED A GIS-BASED WORLD WIDE WEB SITE TO ALLOW THE PUBLIC TO HELP CHOOSE WHICH RUNWAY EXTENSION ALTERNATIVES WORK BEST.

Erie's Internet GIS application is an analytical and public-involvement tool developed for an Environmental Assessment (EA) for the Proposed Runway 6-24 Extension at the Erie International Airport. The system is designed to help the public better understand the proposed project as well as its potential environmental and socio-economic effects. The process transcends the GIS from a tool for planners, managers and experts to its new function as a tool that enhances direct public participation for environmental decision making.

The basic elements of the Internet GIS reflect the contents of a standard EA, which include the following:

- Describe proposed alternatives.
- Present environmental feature maps within the project's area of influence.

- Delineate environmentally sensitive areas and present environmental consequences.
- Incorporate mitigation measures and action plans.
- The Internet GIS is consistent with Federal Aviation Administration (FAA) environmental guidelines, and it promotes the achievement of aviation-related environmental goals, including the following:
- Heighten objectivity in the EA process.
- Provide public information and input.
- Develop a more place-based approach to decision making.
- Ensure that the EA follows an iterative process.
- Comply with FAA's streamlining efforts.
- Develop new tools for interagency coordination.
- Create an effective platform for monitoring and environmental management.

The EA is developed through interagency coordination between the FAA and the Federal Highway Administration, which is currently overseeing the preparation of an EA for the relocation of Pow-

ell Ave., a road affected by the runway extension project.

Project Specifics

The Erie Municipal Airport Authority proposed a runway extension for Runway 6-24, the primary runway at Erie International Airport. The existing runway is 6,500 feet long and 150 feet wide. The proposed project would extend the runway 1,900 feet to the northeast. Safety issues are a primary focus of the proposed extension.

Via the Internet-based GIS, community residents can view the proposed project in the context of their environment. Users can identify tax parcels and relate them to the proposed runway extension and existing land use.

The EA process is evaluating the proposed environmental and socio-economic impacts of all the alternatives identified in the "Master Plan" so alternatives are evaluated using technical and cost considerations as well as environmental and socio-cultural considerations.

A Web site (*http://www.erieairportprojects.org*) was developed to facilitate the link among the Internet mapping system and other related environmental information. The Web site's principal feature is a section on "Environmental Maps," which carries all the features of an Internet GIS. Because the project is developed in conjunction with the Powell Ave. project of the Pennsylvania Department of Transportation, the Internet GIS may likewise be accessed at *http://www.airport-powellprojects.com* under "Project Details, Airport Information."

An Online Mapping System

The main objective for the Internet GIS is to provide public access to information used in developing the EA. Project engineers didn't want to spend much time processing and converting existing data. They also needed a product that could deliver a lightweight Internet application to the public as well as be accessed through a standard and readily accessible Internet client such as a Web browser.

Application Requirements

Features such as runway alternatives, noise contours, runway safety areas and runway protection zones were created in a GIS format. Using a GIS as a data management tool is effective, because it can overlay and query spatial data. Government agencies already had data needed by the project. Tax parcel information was used along with details of the runway alternatives to calculate the number of residential properties affected by the proposed changes. Other information was added to the system, including watersheds, hazardous waste sites, air-quality monitoring data, roads, land use, neighborhoods and socio-economic census data.

By using an aerial photo as a background, a more realistic view of the potential impact areas is provided.

Airport environmental planners considered several different ways to make such information available. If they released the data on a CD-ROM, they would quickly lose control of updates and additions. The project requires data to be centralized, and updates need to be instantly displayed. If the engineers created a kiosk for the public to use at a specific location, they wouldn't be able to service multiple users at one time. An Internet application can handle multiple users from different locations, and the application can be reused and customized to meet the needs of future projects.

From the public's perspective, there are many issues to consider. The application should be quick, easy and convenient to use for someone inexperienced with computers and mapping applications. The project's Internet GIS doesn't require prior GIS knowledge, and there's no need to download or buy extra software. There are different choices for the type of viewer. Some require Active X or Java plug-ins, while others use basic HTML and Java Script, which are lightweight and provide access to all levels of Web users.

The main Web site serves as a platform and provides additional information about the project and the Internet GIS.

Content is another consideration. The site can be used to re-create and investigate maps seen in public meetings. It can be viewed by the public or accessed by people working on the project if they need quick information. The citizens living in the neighborhoods surrounding the airport are interested in evaluating the effect of runway extensions relative to their homes.

In addition, the Internet GIS will give insight to questions such as:

- Where are the extensions proposed?
- Will I have to be relocated?
- Will this change the noise levels around my home?
- Will I be able to receive benefits from the Sound Insulation Program?

Internet GIS Configuration

The tool chosen to connect the public with the data was ESRI Inc.'s ArcIMS software, because most of the base maps and GIS layers were in ESRI's Shapefile format. The specific application directly accepts data that already have been collected and managed for other stages of the runway extension project.

There are three types of data presently included: 1) existing conditions, 2) potential impact areas and 3) cumulative effects. The existing conditions include all existing information such as roads, buildings, wetlands, floodplains, census block data, tax parcel information and airport pavement. Potential impact areas are created for each of the runway alternatives. The primary impact area is the predicted soil disturbance area and a buffer of approximately 500 feet around

the runway safety area. The GIS also takes into account the runway protection zones defined for each alternative. Some cumulative areas will be affected by more than one source, such as a neighborhood block, for example, which will receive a significant increase in noise level and a reduction in air quality.

ArcIMS is a server-side software product that depends on a Web server and a Java servlet engine. Airport engineers already had a Microsoft Windows 2000 server with IIS 5.0 installed to serve company Web pages. There was enough room on the server to install ArcIMS, and the engineers networked another computer to serve the data. They installed The Apache Software Foundation's Jakarta Tomcat 3.2 product as a servlet engine as well as the ArcIMS application and spatial servers. The ArcIMS manager is installed on a separate computer used for development.

When a user views the Web site and sends a request for a map, the Web server sends the request through the Java servlet to the application server, which decides what to do with the request and sends it to the appropriate spatial server to handle the GIS computation. The spatial server then sends an output image of a new map to the user's Web server.

Web Site Design

Citizens need to see more than a map when they visit the Web site. Airport engineers decided to create a site that will introduce the project and EA process. This site includes sections such as "About the Project," "Questions and Answers," "Completed Environmental Reports," "Environmental Maps," "Other Erie International Airport Projects," "News," "Glossary," "Links" and "Public Comments."

The "Environmental Maps" section provides an area where prepared static maps can be viewed using Adobe Acrobat. In addition, the metadata for the layers used in the GIS application are available.

A simple HTML viewer was chosen for the application, because it didn't require users to install plug-ins. It's also the most lightweight in terms of processing required by a client's computer. All the spatial processing is done on the server side, and images are returned to the client. The viewer is interactive—it's able to accept requests and send responses back to clients. Tools available for users include zoom, pan, identify, overview map, view legend, layer control (on and off) and print.

All the layers redisplay images according to user preferences. The setup also allows users to adjust detail levels. The application provides more functions, but airport engineers decided to keep the format simple and only display the necessary options.

Project Concerns

Technical considerations are important when designing an Internet GIS. What type of audience is involved? What type of browsers will citizens have? Do they need an intuitive design or something more advanced? What type of equipment is available?

It's important to determine who is going to host the Web site. The airport project is served from two different locations. An Internet provider hosts the main site that includes the project descriptions, and the Internet GIS application is installed and hosted on a Web server at the C&S office.

The setup for the Internet GIS needs to be carefully planned before installation, and the existing infrastructure should be reviewed. It's important that

network administrators work closely with GIS personnel to design an installation.

A significant issue that airport engineers faced when configuring their software was working around a firewall between the Web server and the rest of the C&S network. Due to some complications with strict firewall settings, the engineers decided to install all the ArcIMS components and data outside the firewall. Typically, this isn't the best installation, and the engineers are currently looking into alternative ways to configure the software.

Lessons Learned

The Internet GIS recently has been completed and released for public use in the EA process of the proposed runway extension. Although positive responses have been received to date, results from the participation effort still aren't available. A few implementation lessons, however, have become apparent.

For example, Internet GIS is a new tool for public involvement in airport development. As such, there's a need to stir more enthusiasm among airport regulators and environmental specialists. Airport engineers also experienced that, unlike a standard GIS, the political concerns are more pronounced in the use of Internet GIS, perhaps because informa-

tion is made more available to the public. Therefore, strategic planning and quality control for any Internet GIS effort are extremely vital to effective implementation.

Map requests are sent from the client to the server, where the Web Server and ArcIMS applications handle the request. The request's result is an output image, which is transferred back to the client.

Internet GIS isn't intended to replace other formats of public participation. It's an approach that complements existing formats, because it allows citizens to access and interact with mapping and GIS data to enhance their knowledge about proposed land development projects in their community and increase their participation in the overall environmental decision-making process.

Internet GIS revolutionizes the way environmental assessment is conducted. As a result, it empowers citizens by providing them with a more dynamic and interactive tool for improved participation.

Calinao is senior environmental planner, C&S Engineers Inc.; e-mail: bcalinao@cscos.com.

Brennan is a GIS specialist, C&S Engineers Inc.; e-mail: cbrennan@cscos.com.

The Future of Imagery and GIS

By Adena Schutzberg

The highlight of the 2003 New England GITA (NEGITA) fall meeting "The Use of Imagery with GIS" was a panel discussion, moderated by Gerry Reymore of Early Endeavors, titled, "The Future of Imagery and GIS." The panelists included Gerry Kinn of Applanix, which is owned by Trimble, Ray Corson of James W. Sewell Company, and Gerald Arp who was with Space Imaging at the time and is now with Booz Allen Hamilton. Kinn provided an introduction by highlighting three trends in airborne imagery.

First, he noted the emergence of "designer" sensors. These are sensors geared to capture very specific types of data, in contrast to the more broadly used imaging and thermal sensors now used on planes and satellites. As an example, he cited a sensor under development at Rochester Institute of Technology that will detect very small fires. Such sensors will become more and more popular in the next five to ten years, Kinn argued.

A second trend Kinn termed "direct geopositioning," essentially the ability to create engineering level accuracy without ground control. Trimble is currently developing such technology. One group that might find such a technology disconcerting is surveyors. But Kinn was quick to point out that surveyors will always have the upper hand since they have the legal authority to interpret and use such data, just as today they interpret data from other sources, such as total stations and GPS receivers.

The third trend is the pent-up demand for GIS data. The graph illustrates a fast growing GIS marketplace. Growth, by Kinn's numbers, runs about 30% per year. On the other hand, remote sensing growth is just a bit better than flat. The gap between the two is pent up demand for GIS data that remote sensing is not

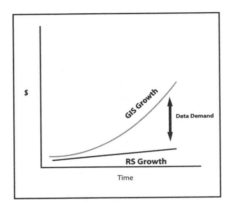

The gap between the growth of GIS and that of remote sensing can be considered the pent-up demand for GIS data that remote sensing is not yet filling.

filling. What other technologies are filling that gap? Other data capture technologies, particularly GPS, says Kinn. The remote sensing community is not delivering, due in part to the lack of tools for automated extraction of features (buildings, impervious surfaces, etc.) from imagery, he argues.

That brought us to the questions for the panel:

1 What is the impact of federal mandates and activities on remote sensing?

ARP: The government has driven the commercial satellite remote sensing with the ClearView and more recent NextView contracts. These are essentially ways for the commercial sector to "fill in" for a dearth of spy satellites. The government is not severely limiting resolution as it's already granted licenses for ½ meter satellite images and made provisions for ¼ meter coverage. The government has also made

it clear that it's trying to avoid sole source contracts (that is, providing a contract to a single vendor without competition) but prefers competition, which provides better quality, coverage, and price.

That said, despite the Open Skies policy, which allows freedom to capture data wherever one can fly, there are some government restrictions on what kind of data may be sold to whom (the denied persons list). For example, Space Imaging can't sell imagery with less than 2-meter resolution of Israel. And, the company can't sell imagery to Saddam Hussein and his peers. Interestingly, imagery vendors can provide data, for instance, of Cuba to Cuba.

KINN: Airborne imaging is like the "wild west" in comparison to satellite imagery. The latter receives government funding and has to obey lots of restrictions from government and nature. The airborne companies, in contrast, have limited funding: a million dollars is quite a lot of development money while satellite folks receive several million at a pop. We are "grubbing the low end." But, we do have some definite advantages: we can land, fix, or replace a sensor, develop a designer sensor, and we need not worry about Kepler's Laws of Motion or even high altitude clouds.

NGA [National Geospatial-Intelligence Agency, which was up until recently known as the National Imagery and Mapping Agency] is actively looking at airborne imaging to augment satellites, though that work is not as heavily funded. Still, there are some quick turn around solutions that should be available soon that will turn heads. For example, technology that allows the delivery of orthophotos as the plane lands may be available in the next year or year-and-a-half.

2 How do privacy issues play into imaging?

ARP: It's odd that privacy concerns are always raised when discussing satellite data, though aerial imagery has far greater resolutions and potential for capturing data that citizens might consider private. Satellite imagery is outside the threshold of personal privacy. Users, public and private, say they want higher and higher resolution, but in fact, few of the imagery purchasers want to process or receive that much data.

CORSON: We have not seen any problems with file sizes and resolutions. In fact, most of our clients are requiring higher pixel resolutions in an attempt to see more detail. Storage space has become inexpensive, making greater resolutions less of an issue.

KINN: It's not so much that we don't want to process or deliver the data; computers are fast and hard drives inexpensive. Instead the issue is the shear amount of raw data turned over to the agencies. As resolution increases, data size goes up as the square, so we are talking about huge data assets. The agencies can't go through it in a meaningful way.

As for privacy, Emerge did a lot of agricultural data capture. Those whose fields were included often asked if we'd be "sharing the data." They were concerned that they might get into trouble with the government since they were growing something they should not. They were particularly concerned that we were making a permanent record of their actions. Privacy goes out the window when you break the law.

3 What changes will we see in delivery—both methods and formats?

ARP: CDs, DVDs, and hard disks won't be going away too soon because Web-based delivery systems still have difficulties delivering very large data sets in reasonable periods of time. We've pretty much standardized on GeoTIFF for delivery. And, most of the government's special formats are losing favor.

KINN: Expect JPEG2000 to be a player soon.

4 With the world as it is, should we expect more government intervention in the coming months and years?

ARP: Actually, we should expect the opposite, as the government's already granting ½ and ¼ meter licenses. The government seems to have concluded that the benefits outweigh the liabilities. That may well be why the Iraqis put their materials underground.

CORSON: There are other ways to make data available but less useful for terrorists. It's possible, for example, to retain relative accuracy, but "play with" absolute accuracy.

5 Does the government have "better" imaging systems than the private sector?

ARP: I don't know and if I did I couldn't tell you. That said, TV and movies do exaggerate quite a bit. Thanks to Hollywood, we get calls from local police departments asking for an image of a particular street on a particular day, hoping we "caught" a robber running down the street. Besides, if someone wants higher resolution than we can provide, an airplane can usually do far better, assuming it can safely reach the geography of interest. One must presume that if the government has licensed commercial satellite companies to go to half-meter resolution, then their systems are still even better.

6 What should we expect regarding data packaging, delivery and licensing?

KINN: The aerial imagery marketplace is quite fragmented. It's tough to make "shrink wrapped boxed products" and sell them to many customers. What we do, essentially, is create custom products. And, to date, no one has figured out how to make 10 cents on each use of an image. We need to do what Kodak did with those little yellow boxes of film: make them ubiquitous and make a profit on each one.

As the entire imagery market matures, we are more likely to see a few bigger firms providing consumer level products/ services. So, how will a future imagery company make money? Say sometime in the future you go to Wal-Mart to buy grass seed and fertilizer. New regulations limit the amount of fertilizer you can use on your property to cut water pollution. The salesperson uses a Web service to check out the size of your lawn via an image, then provides you with the right amount. You pay no more, but Wal-Mart sends a few pennies to the imagery company.

As for licensing, aerial imagery has a shelf life similar to bread. It's very tasty when fresh, but less and less interesting as it ages. Licensing will change to take that into consideration, but licensing will not go away.

ARP: Our licensing in the past was quite challenging. We've changed it accordingly so that our licensing is much simpler for "new" data. By the way, our images are already at Wal-Mart, in some of the video games. The new licensing takes advantage of the fact that in a few years the game maker will come back to us for up-to-date imagery for the cities covered in the games. In general our licensing is less and less restrictive, especially with the government customer.

CORSON: We actually have a historical film archive of many of the areas we've flown over the years dating back to the 1950s. A large percentage of our clients are interested in historical photography and mapping. The big challenge for us is being able to access this information digitally. We are scanning the archive film and creating metadata for it so we can find the images of interest in a massive project.

7 How will "designer sensors" play out?

KINN: Digital imagery is not being driven by remote sensing, but rather by other uses, such as taking pictures of Aunt Martha. What's happening in our corner is taking the off-the-shelf results of advances in other areas, and bolting them together to create "exotic" sensors. These might be used to sense specific targets like the ones currently in development at the Rochester Institute of Technology to sense tiny fires. Or, we might use multiple cameras to mimic the coverage and resolution of a single large "chip" that may be very expensive or not even exist. At some point sensors may be specialized enough to fly one mission and be retired. Of course, that's possible in aerial imaging, but not on satellites.

ARP: I don't agree with the designer view. In the commercial area the only change is in resolution, there is not a large push to add bands. To that end it is interesting to note that the latest satellite out of India, the P-6, [also known as ResourceSat] launched in October 2003 includes a green, red, and infrared band, but no blue! And there's no hyperspectral as there is not enough demand for it; it's too specialized for a commercial satellite.

The commercial market, which for us means primarily the government, is pushing exactly this type of trend.

8 How will LiDAR play into imaging?

KINN: Unlike imagery of a city, 95% of a DEM doesn't move over time. So, LiDAR doesn't require the refresh rate needed for say municipal applications. That said, "exotic" uses for LiDAR are on the horizon such as measuring crop height and canopy depth.

9 What can we expect when we need coverage of say a county, on demand?

ARP: Back a few years that was quite a challenging task. Over the years we've learned to "fly" the satellite and deliver those types of products. We are getting much better at that type of work.

10 What's the future of imagery prices?

KINN: Prices are certainly coming down. Other markets for "high tech" products suggest that specialized uses of technology give way in time to commercial uses and ultimately to consumer uses, with a corresponding drop in price. Consider what has happened with computers, for example. In GIS, we are dropping down to consumer products in services like MapQuest, but the business model for remote sensing is still unclear.

ARP: I agree that prices are coming down, but products will remain primarily custom. A few years ago we offered digital ortho quads. We expected to make a killing from people who wanted a picture of their house. Bottom line: it was a big yawn. We haven't found that niche. In point of fact, even our "standard products" are still custom.

KINN: The big issue is still that we offer data, raw data. We've not figured out how to offer the answer to the question, that is, information. That's why I think we keep coming back to the idea of automated extraction as a key tool to open the market. It can help change this raw data into information.

About the Author
Adena Schutzberg is editor of GIS Monitor, *a weekly e-mail newsletter from* GITC America, Inc. *She runs ABS Consulting Group, Inc. in Somerville, Massachusetts.*

From *Earth Observation Magazine,* February/March 2004, pp. 29-32. Copyright © 2004 by GITC America, Inc. Reprinted by permission.

Calling All Nations

Globalization has produced an explosion in international phone traffic

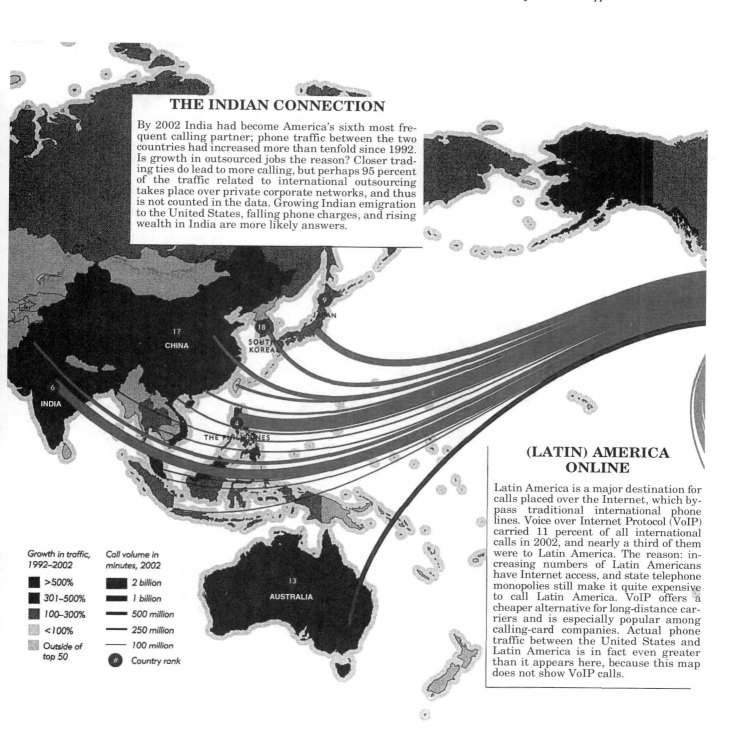

THE INDIAN CONNECTION

By 2002 India had become America's sixth most frequent calling partner; phone traffic between the two countries had increased more than tenfold since 1992. Is growth in outsourced jobs the reason? Closer trading ties do lead to more calling, but perhaps 95 percent of the traffic related to international outsourcing takes place over private corporate networks, and thus is not counted in the data. Growing Indian emigration to the United States, falling phone charges, and rising wealth in India are more likely answers.

(LATIN) AMERICA ONLINE

Latin America is a major destination for calls placed over the Internet, which bypass traditional international phone lines. Voice over Internet Protocol (VoIP) carried 11 percent of all international calls in 2002, and nearly a third of them were to Latin America. The reason: increasing numbers of Latin Americans have Internet access, and state telephone monopolies still make it quite expensive to call Latin America. VoIP offers a cheaper alternative for long-distance carriers and is especially popular among calling-card companies. Actual phone traffic between the United States and Latin America is in fact even greater than it appears here, because this map does not show VoIP calls.

Growth in traffic, 1992–2002

- >500%
- 301–500%
- 100–300%
- <100%
- Outside of top 50

Call volume in minutes, 2002

- 2 billion
- 1 billion
- 500 million
- 250 million
- 100 million
- # Country rank

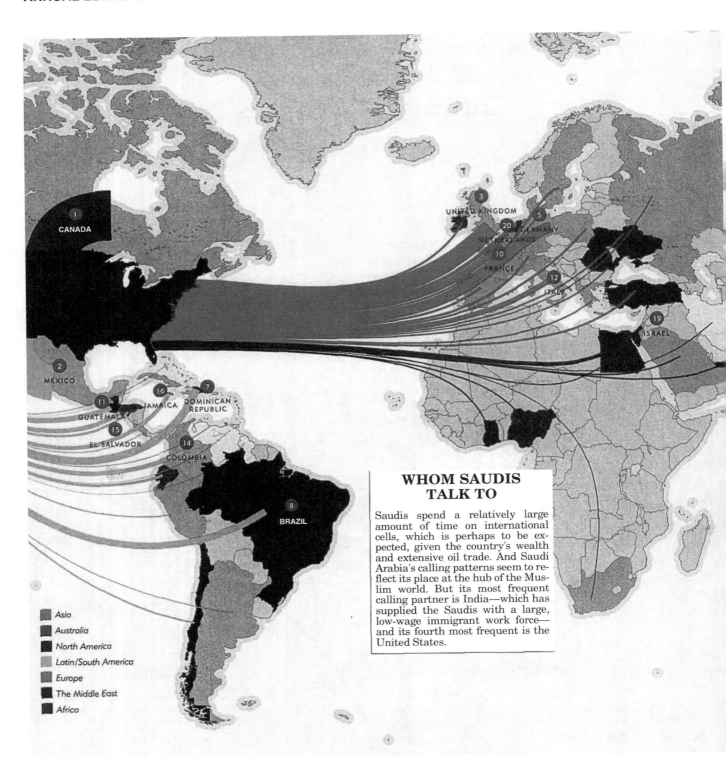

WHOM SAUDIS TALK TO

Saudis spend a relatively large amount of time on international cells, which is perhaps to be expected, given the country's wealth and extensive oil trade. And Saudi Arabia's calling patterns seem to reflect its place at the hub of the Muslim world. But its most frequent calling partner is India—which has supplied the Saudis with a large, low-wage immigrant work force—and its fourth most frequent is the United States.

Asia
Australia
North America
Latin/South America
Europe
The Middle East
Africa

Americans are sometimes accused of being isolationist, culturally incurious, and ignorant of world affairs. All this may be true—at least to some extent. But we do spend an awful lot of time talking to people around the world. In 2002 international telephone traffic between the United States and other countries reached 54.6 billion minutes. There is now more phone traffic each year between the United States and the Philippines, to cite just one example, than between the UK and France, or France and Germany. And although many Europeans talk extensively to one another across national borders, the United States is alone in the breadth of its outreach: of the 116 intercontinental routes that each carried 100 million minutes or more from one country to another in 2002, seventy-six involved the United States.

The volume of phone traffic into and out of this country reflects its prominence as both a trading partner and a destination for immigrants. The map above shows traffic between America and its fifty most frequent calling partners, with the bands widening as volume increases. As one would expect, thick connections link the United States to Europe; just a decade ago Europe was our largest calling partner by a substantial margin. But the connection to Latin America is now significantly thicker, and the connection to Asia is nearly as thick. Only a few tendrils reach into Africa and the Middle East—regions that in many senses remain culturally disconnected from the United States.

Growth in international calling since the early 1990s illustrates the startling pace of globalization. Total phone traffic between the United States and its fifty most frequent calling partners grew by 362 percent from 1992 to 2002, according to data collected by TeleGeography, a consulting firm for the telecommunications industry. Over the same period the volume of calls between the United States and Brazil, China, and India grew by 562 percent, 649 percent, and 1,028 percent, respectively.

What accounts for this explosive growth? The cost of international calls has declined dramatically as many countries have dismantled telecom monopolies, subjecting phone carriers to greater competition. And anybody can now put down a little cash and pick up an international calling card or mobile phone. In this country recent immigrants have used prepaid calling cards to overcome credit and language barriers that once shut them out of international phone service (it's easier to get a card at the corner store than to deal with the phone company). As a result, the map of America's international calling patterns might easily be mistaken for a map of the immigration routes that brought us here.

—NATHAN LITTLEFIELD

MAPPING THE NATURE OF DIVERSITY

A LANDMARK PROJECT REVEALS A REMARKABLE CORRESPONDENCE BETWEEN INDIGENOUS LAND USE AND THE SURVIVAL OF NATURAL AREAS

Maps may be famously variable in accuracy, but generally speaking they are no more "objective" than are movies, novels, speeches, or paintings. Even if painstakingly accurate, they heavily reflect the interests of those who paid to have them made. Those interests may be political, commercial, or scientific. In the second half of the twentieth century, world maps emphasized the preoccupations of the Cold War, with a primary emphasis on international borders. The globes we had in our classrooms showed a world made up of nations. Until recently, most maps showed very little of what some of us now believe to be critical to the future of life: the boundaries of bioregions, watersheds, forests, ice caps, and biodiversity hotspots—and the principal ocean currents, wind currents, oceanic fisheries, and migratory flyways. In one of the offices at Worldwatch, there's a large map of North America showing nothing but the distribution of underground water. In *World Watch*, over the years, we've published maps of the global distribution of infectious diseases, war, slavery, refugee flows, and electric light as seen from space. The advancing technologies of Geographic Information Systems (GIS), combining the use of satellite imaging and digital data, have made these tasks easier by replacing laborious cartographic handwork with a capacity to superimpose maps of various elements showing how these elements may be related.

. . . [seeWorld Watch, March/April 2003, for a fold-out map that was] created under the direction of a nonprofit group called the Center for the Support of Native Lands, and produced in its final form by the National Geographic Society. [The map] was designed to exhibit two main categories of information: the distribution of cultural diversity in Central America and southern Mexico, and the distribution of forest and marine resources in that region. By superimposing these sets of information in detail, the map strongly confirms a hypothesis that has long

been familiar to environmentalists and anthropologists alike: that there is a significant correlation of some kind between cultural diversity and biological diversity. That may seem obvious, as the homogenizing impacts of globalization are fueled and further exacerbated by a stripping of forests for cattle-ranching, plantations, and urban development. But in the past, the kind of data available to demonstrate this correlation on a regional or global basis has been fairly broad-brush. In 1992, for example, Worldwatch published a paper by Alan Durning, *Guardians of the Land: Indigenous Peoples and the Health of the Earth*, which included a diagram showing which nations had the highest cultural diversity (defined as those in which more than 200 languages are spoken) and which had the highest biological diversity (those with the highest numbers of unique species). Of the nine countries with the highest cultural diversity, six also ranked among those with the highest numbers of endemic species.

The history of Native Land's map of Central America and southern Mexico can be traced back even further, to the publication of the book *Regions of Refuge*, by the Mexican anthropologist Gonzalo Aguirre Beltrán, in 1967. Aguirre Beltrán noted that beginning with the Spanish conquest of Mesoamerica in the sixteenth century, indigenous peoples who had been decimated by warfare and by diseases against which they had no resistance sought refuge in "particularly hostile landscapes or areas of difficult access to human circulation." It was in those remote, often mountainous or jungle-covered areas that the refugees were able to rebuild their societies and preserve their cultures—and it is in those areas that they survive today.

In 1991, anthropologist Mac Chapin, who was then working for the Central America program of Cultural Survival, found himself perusing a map entitled "Indians of Central America 1980s," which had been compiled by the Louisiana State University Department of Geography and

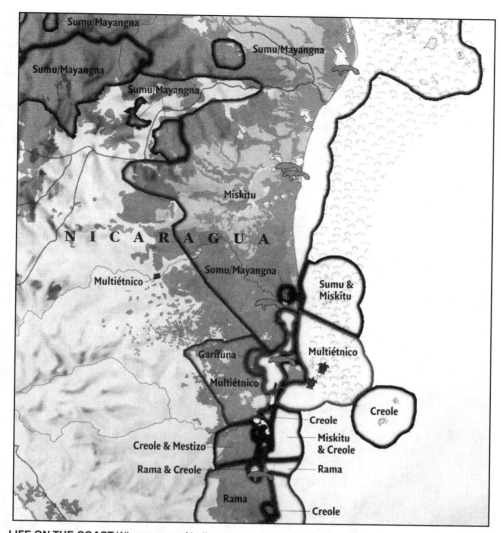

Sumu/Mayangna

Sumu/Mayangna

Sumu/Mayangna

Sumu/Mayangna

Sumu/Mayangna

Miskitu

N I C A R A G U A

Sumu/Mayangna

Sumu & Miskitu

Multiétnico

Multiétnico

Garífuna

Multiétnico

Creole

Creole

Creole & Mestizo

Miskitu & Creole

Rama & Creole

Rama

Rama

Creole

LIFE ON THE COAST When a map of indigenous territories (outlined areas) is superimposed on a map of forest cover (darker areas), the correspondence is close—as seen in this section of the Caribbean coast of Nicaragua. One reason for this correspondence is historic; the native groups retreated to densely forested areas centuries ago to avoid extermination by conquistadors. Another reason may be ecological, as the Indians' subsistence economy has proved less destructive to natural resources than has the "developed" economy. The ecological interdependence of forests and waters (estuaries delivering nutrients to the water, coastal land providing marine turtle nesting sites, etc.) is reflected in the mapping of indigenous territory, which is as much marine as terrestrial.

Anthropology two years earlier. He noticed that the indigenous population was located primarily in two areas—in highland Guatemala and strung out like a chain of beads along the Caribbean coast. "As I took this in," Chapin recalls, "I periodically looked at a 1986 National Geographic map of Central America hanging on the wall before me." The primary display was a standard "political" map, but in the corner was a small inset map showing, somewhat crudely, the region's vegetation. According to this map, most of Central America's natural forest cover was to be found hugging the Caribbean side of the isthmus—precisely where the lowlands indigenous peoples lived.

Chapin began thinking about the possibility of making a map showing the correspondence of indigenous settlement and forest cover, and shortly thereafter he was invited by Anthony de Souza, editor of the National

Geographic journal *Research & Exploration*, to make one. The map—a precursor to the one which appears here [see "Life on the Coast"]—was published in that journal in 1992. It had small circulation but huge impact. Copies ended up on walls at the Inter-American Development Bank, World Bank, UN Food and Agricultural Organization, and the private residence of the president of Guatemala. One of the strongest impacts was on the indigenous peoples of the region. "The map helped to strengthen what soon became a widespread campaign for protecting and legalizing their territories," says Chapin.

That campaign also opened up a new venture for Native Lands—helping indigenous groups document their own land-use and marine-use patterns for purposes offending of incursions by developers, squatters, loggers, and the like. Traditionally, most of the native communities had regarded their territories as commons, and had never seen any need for such documents as plats and deeds; but the lack of such "proof" of ownership meant they were often unable to defend their territories from being occupied or exploited by outsiders. Chapin and his colleagues, who by 1994 had left Cultural Survival for Native Lands, embarked on a series of "participatory mapping" projects, in which indigenous groups made hand-drawn maps of their ancestral lands, and these maps were then combined with inputs from aerial photograph interpreters and cartographers to produce highly detailed, small-scale maps of the tribal territories. (One of the hand-drawn maps was published on the back cover of the January/February 1994 *World Watch*.) The progress of the participatory mapping program is summarized in a book coauthored by Chapin and his colleague Bill Threlkeld, *Indigenous Landscapes: A Study in Ethnocartography*, published in 2001.

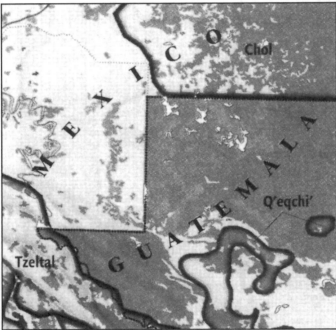

NATIONAL POLICIES, as well as cultural practices, can make a large difference in protection of natural assets. In this view, the border between Mexico and Guatemala demarcates a stark contrast between the land Mexico has allowed to be cleared for cattle grazing or timber, and the intact forest which remains across the border to the south and east.

In 2000, Native Lands decided to do an update of the original Central America map. "We knew that deforestation had advanced and that problems with Central America's Caribbean coastal environment—the bleaching of coral reefs and decline in fisheries—were increasing." Several heavily financed conservation projects had failed to halt the destruction. (The Mesoamerican Biological Corridor program of the World Bank had spent close to

$100 million in the 1990s, and according to several assessments had little to show for it.)

It was also possible, by 2000, to make a far more accurate and informative map than the original. There had been major advances in satellite technology, and the success of the first version mobilized greater financial and professional resources. The original map had not included features of marine ecosystems, and had shown indigenous territories only on land. The new one would show marine-use areas, which are of critical importance to indigenous peoples along the Caribbean coast. The new map would also include the Maya region of southern Mexico, which the first map had not. The project would integrate the work of indigenous representatives, anthropologists, ecologists, and cartographers from every country from Mexico to Panama.

FROM THE MAP...

The first human footprints appeared along the isthmus that is today Central America and southern Mexico as early as 18,000 years ago. They were made by small bands of hunters and gatherers moving south through pristine landscapes abundant with plant and animal life. The newcomers prospered. They put down roots, and spread out, adapting themselves to the region's varied ecosystems. When the Europeans arrived at the end of 15th century, the native population was an estimated 7.680,000 people who spoke at least 62 languages and had cultural configurations ranging from tiny foraging tribes to the complex civilization of the Maya.

Contact proved disastrous for the indigenous peoples. As many as 90 percent of them died within the first 100 years, mainly from diseases against which they had no resistance. Most of those who survived retreated into the hinterlands to escape the unseen pestilence holding out in the remote fastness of the northern highlands and the humid forests of the Caribbean coastal slope. Indigenous peoples still have a strong presence in these areas today. Their numbers have steadily increased and now surpass pre-Hispanic levels with some 11 million people arrayed among more than 60 ethnic/linguistic groups. They are currently mounting campaigns to protect their ancestral homelands, natural resources and distinctive cultures.

Data-gathering took 15 months. The map was then designed by National Geographic Maps, the cartographic division of the National Geographic Society, and printed in an oversized (44 x 27-inch) format. The total cost, including data-gathering, design, and production, came to about $400,000. The map was printed with Spanish and English text and published as an insert to the February 2003 issue of the Latin American edition of National Geographic magazine, *National Geographic en Español*, with about 130,000 copies to be distributed in Central America and Mexico.

—*Ed Ayres*

From *World Watch*, Vol. 16, No. 2, March/April 2003, pp. 30-32. © 2003 by Worldwatch Institute, www.worldwatch.org.

MAPPING

Fortune Teller

For magazine readers of the 1940s, architect-turned-artist Richard Harrison mapped their world in a bold new way.

BY ANN DE FOREST

"THREE APPROACHES TO THE U.S.," READS THE HEADLINE. Underneath, printed in vivid color on a page of the old magazine, are three maps, three rounded edges of the globe, overlapping, like an illustration of a planet rising. In each, the land spreads out tan and yellow against the blue-green water. It almost seems that you're flying over this contoured terrain, about to swoop down for a landing. But what terrain is it? The headline says it's the United States, but these maps don't look like any America you know. Where is that familiar, iconic national outline? Where is Maine, waving in the upper right-hand corner? And where is California, bending forward like a ship's prow on the left?

The man who made these maps, Richard Edes Harrison, cartographer for *Fortune* for more than 20 years, specialized in such unorthodox presentations. He sought to, in his own words, "jolt... [readers]... with a new and refreshing viewpoint." So, on a Harrison map, north isn't always up, the earth isn't always flattened on the page and perspective—the point from which one views the planet—shifts, depending on what information he's aiming to convey. Says Joanne Perry, maps librarian and head of cartographic services at Pennsylvania State University. "He was trying to put out maps that showed truth, not convention."

In September 1940, the truth these particular maps revealed was far more ominous than refreshing. More than a year before the Japanese attack on Pearl Harbor, *Fortune* featured the images as part of an *Atlas for the U.S. Citizen:* 11 maps designed to show Americans they weren't as isolated from events in Europe and Pacific as they might like to believe. That swooping perspective, so breathtaking today, is actually the enemy's-eye view. Approached by air from Berlin, Tokyo or even Caracas, Venezuela, the United States suddenly looks vulnerable, with Detroit and Chicago tempting targets for the Germans, and Seattle and San Francisco poking perilously in Tokyo's flight path. And "if an enemy...," a caption warns, "should ever establish himself on the northern shore of South America or in the mazes of the West Indies," the Gulf Coast, hub of industry, becomes America's "soft belly."

Harrison, architect by training, artist by inclination, came to mapmaking accidentally (though he'd done some scientific illustrations for pay before studying architecture). Like many Americans in 1932, he needed work. A friend recommended him for an assignment to *Time* to make a map projecting the outcome of the looming presidential election. In 1936, he was kicked upstairs to another magazine in the Henry Luce empire, *Fortune,* where as part of its crack graphics team, he applied his visual skills to explicating the macro—charting the worldwide holdings of General Motors—and the micro—diagramming the inner workings of a gas mask.

World War II proved the ideal subject for his talents. This was a new kind of war, a truly global war, fought in the skies as well as on land and at sea. It was a war that demanded entirely new maps, new ways of seeing the world. Here, Harrison's lack of formal cartographic training was an advantage. Not bound by convention, he could make maps that opened readers' eyes to the realities of global geography in an aviation age. ical tract/advertising pamphlet published by Consolidated Vultee Air-

"He explicitly saw himself as speaking to the general public, while he saw cartographers as failing in this regard," says Susan Schulten, a history professor at the University of Denver and author of *The Geographical Imagination in America, 1880–1950.* She interviewed the "witty, subversive" carto-journalist, then 90 years old, shortly before his death in 1993. Certainly, the wide range of maps he made for *Fortune* in the 1940s show him reveling in his amateur status. He tweaks professional geographers for their staunch commitment to the Mercator projection, the standard flat map of the world ("a dangerous map for world strategy," reads one caption). And he always finds space to explain, through clever illustrations and entertaining examples, exactly why he chose a particular style of mapmaking for a particular purpose. Never condescending, Harrison takes the tone of a fellow layman, and enthusiastic amateur, who just happened to discover what terms like "azimuthal equidistant" meant himself.

His map of "The World Divided," published in August 1941, centers on the North Pole, with the rest of the world a great circle around it. (In a typically charming sidebar, Harrison explains the principle behind this map with two cartoons of a dancer. In the first, she stands at rest, with her skirt as the globe; in the next, she's twirling, with the globe having risen and flattened to a disk.) This world, "divided" into Axis nations, Anti-Axis and various positions in between by color and shading, is also inextricably united: The U.S.S.R. and Alaska touch on this map, while the Aleutian Islands are part of one long, necklace-like chain that extends through Japan and down to the Philippines.

This view was prescient, of course. In March 1942, a nearly identical "polar equidistant" map appeared in *Fortune,* titled "One World, One War." Indeed, once the United States entered the war, Harrison's maps became integral to *Fortune*'s reports on U.S. action and changing strategic situations worldwide. The sweeping, slightly rounded views gave "a sense of direction, a sense of the movement of the war," says Schulten.

Harrison's work also serves to underscore the advances in air navigation at the time. As Schulten wrote in her book, "The use of a polar route to connect Japan to Alaska effectively transformed the Pacific from a massive body of water protecting the United States into a smallish lake."

In 1942, *Fortune* printed a stunning series of colorful, fold-out maps of the Pacific, Atlantic and Arctic arenas. These, says Perry, "make you see why something is happening in the world: Why are our troops in North Africa?... Why are your sons and neighbors dying in different places?" His vivid creations were phenomenally popular, reprinted by the military and various airlines and displayed at post offices.

Capitalizing on the cartographer's celebrity, *Fortune* published a book of Harrison's wartime maps in 1944. *Look at the World: The Fortune Atlas for World Strategy* was an instant best seller. Immediately after World War II, Harrison collaborated on numerous books like *Compass of the World: A Symposium on Political Geography* and *Maps and How to Understand Them,* an unusual hybrid textbook/politcraft Corporation promoting "Air Supremacy—For Enduring Peace."

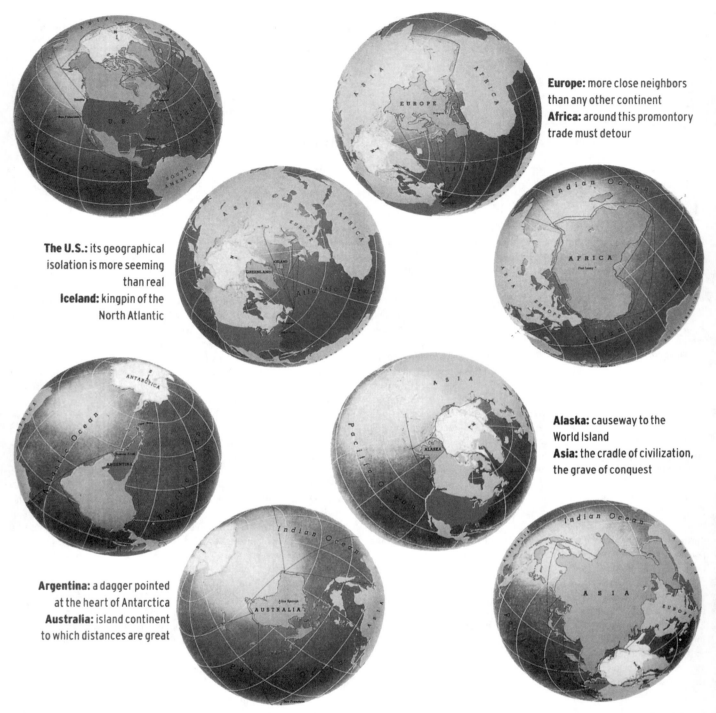

Europe: more close neighbors than any other continent
Africa: around this promontory trade must detour

The U.S.: its geographical isolation is more seeming than real
Iceland: kingpin of the North Atlantic

Alaska: causeway to the World Island
Asia: the cradle of civilization, the grave of conquest

Argentina: a dagger pointed at the heart of Antarctica
Australia: island continent to which distances are great

Planet Life Harrison's "Eight Views of the World" was created for *Look at the World*. The captions are from the original. Used with permission of Ross & Harrison III.

Today, though, *Fortune*'s "celebrated cartographer" is all but forgotten. The Mercator projection—that flat, 433-year-old map designed for an age of navigation, not aviation—has prevailed. "Perhaps his style of maps were really keyed into a time of crisis," Schulten speculates. "He had an advantage in the 1940s. While Rand McNally had to create maps that would last," Harrison, working for a monthly magazine, was making maps that were newsworthy. "He was drawing not just for the war, but for that week of the war."

We all know the fate of yesterday's news. It's too bad, though. Harrison's fresh-eyed perspectives and projections deserve a revival. Crisis or not, when it comes to viewing the world, we can always stand a "jolt… with a new and refreshing viewpoint."

Speaking of fortune, Navigator *is fortunate to have* ANN DE FOREST *as its resident cartographile.*

Asphalt
and the Jungle

A road project in the Amazon may be the world's boldest attempt to reconcile growth and conservation

YOU do not drive on the right of the BR-163, nor do you drive on the left. You drive on whichever bit of the road seems least likely to tear off the undercarriage of your vehicle. During the six-month rainy season, when the road becomes a river of mud, men with tractors wait for you to founder and haul you out for a fee. Under such conditions, the 1,765km (1,097 mile) journey from Santarém, a port on the Amazon River, to Cuiabà, capital of the state of Mato Grosso, can take a fortnight.

Within four years, if Brazil's government has its way, the BR-163 will be a super-highway, launching commodities towards markets in Europe and Asia, speeding computers and cell phones from Manaus to São Paulo and ending the near-isolation of hundreds of thousands of people living along its unpaved stretches.

Yet the paving of the BR-163 is feared as much as it is yearned for. The road joins what Brazilians call, without great exaggeration, the "world's breadbasket" to the "world's lungs"—the fields and pastures of Mato Grosso to the Amazonian rainforest. If the past is any guide, the lungs will suffer. Paving the BR-163 could lay waste to thousands of square kilometres of forest, carrying deep into the jungle the "arc of deforestation" through which it passes. It may visit similar destruction on the small farmers, gatherers and indigenous folk clustered along its axis. In Parà, the more northerly of the BR-163 states, older settlers are already battling loggers and land grabbers up and down the road. "On the one hand [it] will bring development," says Cícero Pereira da Silva Oliveira, head of the union of rural workers in Trairão, a settlement 380km south-west of Santarém. "On the other it will bring ruin to the region—more land grabbing, more drug trafficking. Total violence will arrive."

That would be a local disaster with global implications. During the 1990s, deforestation may have accounted for 10-20% of the carbon released into the atmosphere. Road development could deforest 30-40% of the Amazon by 2020, according to one estimate. But the paving of the BR–163 is supposed to be a different sort of roadworks, bringing growth that is ordered rather than chaotic, reducing social inequities rather than exacerbating them, preserving the Amazon rather than despoiling it.

Getting it right has now become a global project, involving NGOs, multinationals and grass-roots groups, as well as all levels of Brazil's government. There are plenty of disagreements, but this throng is forming unlikely alliances, overturning assumptions about how to police the forest and proposing novel ideas for reconciling growth and conservation.

The road was opened 30 years ago by dictators whose idea of manifest destiny was to send bulldozers to clear a trail into the forest and entice people to follow with the prospect of land and subsidies. "Land without people for people without land," they urged, and many responded, settling along the margins of thoroughfares that took turns as dust and mud. One indigenous tribe, the Panarà, was decimated by viruses brought by the settlers and expelled from its traditional territory. The BR-163 hosts what the transport ministry calls "the highest concentration of slave labour in the known world."

Most governments since have promised to pave it. That the pledge may finally now be redeemed owes less to the demands of those living along the unpaved stretch, which lies mainly in Parà, than to the interests gathered at either end of the Cuiabà-Santarém road.

It starts in the capital city of Mato Grosso, a state that calls itself the "Amazon tiger". While Brazil's economy shrank last year, Mato Grosso's GDP grew 8% thanks to a boom in soya, beef and other commodities. Such exports, notes Blairo Maggi, the state's governor, largely account for Brazil's trade surplus. This shields an indebted economy from chaos.

At the northern terminus, in the subdued port town of Santarém, stands a $20m grain terminal built by Cargill, an American trading company whose logo is now the town's most visible landmark. The terminal is handling grain delivered by river but will really come into its own when the BR-163 is ready for lorries bearing grain from Mato Grosso.

The payoff will be stunning. The farmers of Lucas do Rio Verde currently ship their production out through the congested ports of Santos and Paranaguà in Brazil's southeast. Paving the BR-163 would halve the time and cost of transport, reckons the town's mayor, Otaviano Olavo Pivetta. That would inject 37m *reais* ($12m) into the local economy, a gain that would be repeated across Brazil's central-western region. All Brazilian agriculture, which is already intimidating rivals abroad, will be more competitive.

Manufacturers in the duty-free zone of Manaus, who see the value of their tax breaks eaten away by the cost of delivering their fragile electronic goods via bumpy highways, expect freight costs to fall by 300m *reais*. The paving of the BR-163, which is to be a privately operated toll road, is a big part of the solution to Brazil's *apagão logística*—its logistical blackout—which threatens to choke off an economy that is just beginning to grow again. The road itself makes what looks like an irrefutable argument for an upgrade. In soya-growing Lucas, which lies along the paved stretch, municipal schools have semi-Olympic-sized swimming pools. Trairão, on the other hand, lacks not only asphalt but basic sanitation.

The cost of progress?

The road will transform as well as transport, but not necessarily for the better. The Amazon forest has already shrunk by 15% since the 1960s. In general, some 85% of deforestation takes place within 50km of a road, because a road makes it more profitable to fell trees, first for timber and then for pasture, the biggest contributor to the denuding of the forest. The paving of the BR-163, which passes through one of the Amazon's most varied bird habitats, will destroy 22,000-49,000 square kilometres of forest within 35 years, according to a report in 2002 by two research institutes, IPAM and the Instituto Socioambiental. Without law and order, the road could usher in the strong and flush out the weak.

Trairão is a jangling ten hours by road from Santarém. Small farms and pasture line the verges of the highway, and lorries loaded with *ipé*, a tropical hardwood, ply it ceaselessly. The municipality has ranching on a small scale but Ademar Baú, a farmer who is also Trairão's mayor, sees great possibilities. The region is "very suitable" for cattle, he says, with lots of rain and no disease. Farmers are beginning to experiment with rice and soya.

But the logging is clandestine, and the farming takes place in a legal limbo. Mr Baú says that 90% of the proprietors in Trairão have no clear title to their land. He blames this on "bureaucracy", in particular the federal agrarian-reform agency, called INCRA. Because INCRA rarely sells land outright, and then usually in lots of 100 hectares, landowners acquire it through fronts, cannot borrow money from banks and cannot get official sanction for logging.

With the paving of the road in prospect, Trairão is experiencing a boom. Its population has swelled from 14,000 to 25,000 in the past two years. The price of land along the road has jumped nearly tenfold. So common are overlapping claims, says Mr Baú, that if all were valid Trairão would rise three storeys high.

Like much of Parà, Trairão is caught up in what may be the unruliest property market in the world. The federal government owns 70% of the state's land, through agencies such as INCRA, Indian reserves and national forests. Much of INCRA's property is federal in name only, and thus an invitation to *grilagem*, which refers to an earlier practice of putting land deeds in boxes of crickets to make them look authentically antique.

There are trappings of legitimacy. Enterprises apply to INCRA for documents called *protocolos*, based on parcels of land surveyed from the air and impressively "geo-referenced". These are then sold to loggers, speculators or aspiring ranchers, sometimes via the internet. The *protocolo* concedes no right of ownership, yet is bought and sold as if it does. Sometimes it is enough to persuade earlier settlers to leave. If not, there can be violence. "When I arrive, the *ribeirinho* [river dweller] is there," says a logger from Itaituba, a district along the BR-163. "He's been there 80 years. I have the document. That means a battle, sometimes to the death."

Land clashes, an old story in southern Parà, rage along the BR-163. In Santarém, Cargill's terminal has opened up a new front in Brazil's soyabean boom. By night, planters recently arrived from Mato Grosso cruise the waterfront in shiny Hilux pickup trucks, a marker of rural prosperity. Often, the locals are happy to sell out to deep-pocketed buyers. Sometimes, alleges a local organisation of family farmers, they are pressured to leave. In Castelo dos Sonhos ("Castle of Dreams"), four out of five corpses in the cemetery are those of murdered members of the rural workers' union, says Socorro Pena of IPAM. Over 500 people have died in Parà's land wars.

Development and conservation

Brazil has grown up since the generals etched the BR-163 into the forest. No longer does the government think it sufficient to build something and then abandon it. Now, 15 federal ministries are pondering every conceivable consequence of paving the road, from greater prostitution to opportunities for organic derivatives of castor oil. Social movements, which barely existed 30 years ago, are subjecting the project to intense public grilling. For them, government services, from policing to education, are as important as conservation.

The government envisages a two-phase process to soften the impact of the road: first, "emergency actions" to accompany roadworks starting next year, such as beefing up law enforcement and settling property claims, and then an environmentally friendly master plan for the road's "area of influence"—nearly 1m square kilometres and 1.7m people. "The predatory paradigm of the past 500 years is being broken," boasts Alexandre Gavriloff, the transport ministry's director of concessions.

A traveller cannot help but wonder whether government can break paradigms on its own. The Santarém branch of INCRA administers federal property spread across 17m hectares in 11 districts. Two vehicles are working properly and one is barely functioning, says the unit's chief, Pedro Aquino de Santana. Its employees are mostly too decrepit to travel; the youngest was hired 21 years ago. IBAMA, the federal issuer of environmental licences, is hardly in better shape. Flàvio Montiel, its director of environmental protection, says that "IBAMA today doesn't have the means to supervise" but contends that things are improving. After seeing its budget for policing Brazil's forests slashed to a derisory 17m *reais* this year, it has received an emergency infusion of funds.

The government proposes to ride up the BR-163 like a no-nonsense sheriff on the American frontier, but a truer analogy would be to 19th-century Afghanistan, which drew outside powers and local potentates into a contest for influence. The players of this latter-day Great Game are an assortment of local and global pressure-groups, multinationals, foreign lenders and various levels of government with diverse interests. Yet they have mounted a challenge to the traditional approach to managing the forest, which relies on enforcement, not inducement. "The government is not set up to be able to manage the Amazon," says Michael Jenkins of the Katoomba Group, which marshals private-sector incentives for green ends. Alliances between environmentalists and enterprises can help, he thinks. That means "folks traditionally seen as arch enemies need to be in the same room with us."

No enemy is more arch than Mr Maggi, who, besides being Mato Grosso's governor, is part-owner of the world's biggest soya producer. Anything that is good for soya is bad for Brazil, many Brazilians believe. It contributes to deforestation, usually indirectly, by occupying pasture and pushing ranchers deeper into the forest; it poisons rivers with pesticides. Soya planters amass land but employ few people. During the year in which Mr Maggi took office, Mato Grosso's rate of deforestation more than doubled. No coincidence, his critics said.

Mr Maggi is certainly no tree hugger. Asked about the effects of growth on the environment, he replies that "you don't make an omelette without breaking eggs." He opposes new reserves for indigenous people, who, so far, have been the forest's most reliable protectors. Yet Mr Maggi vows to defend the law, which in Brazil is strict: 80% of densely forested private land may not be cleared, nor may the banks of rivers. As the Amazon links its

economy to that of the rest of the world, the cost of flouting the law is mounting.

The Maggis' company, Grupo André Maggi, is the best example. Its European customers, tutored by pressure-groups, are nervous about the fate of the forest, and its bankers are becoming so. The Maggi company wants to borrow $30m from the International Finance Corporation (IFC), an arm of the World Bank that imposes relatively strict green norms on its clients. It is also a client of Banco Real, owned by a Dutch Bank called ABN-AMRO, one of several summoned by NGOs last November to be warned against financing tree-toppling soya. The IFC named a mediator between the Maggi company and the NGOs. With their encouragement, it obeys the law and insists that the 500 other farmers from which it buys soya do so as well.

NGOs are putting similar pressure on Maggi's competitors: the Nature Conservancy is tackling Cargill, for example. The ultimate aim is a system of certification, assuring consumers in the first world that soya growers are complying with environmental law. It is harder—and more important—to extend that idea to the frontier, where anonymous ranchers are doing most of the deforesting, but there is progress here, too. Brascan, a Canadian-owned company, may soon start up ranching in the Amazon; Maggi-like, it will lend cattle to other ranchers provided they obey the law.

Private schemes underpin public policies that are themselves in flux. Pretty much everyone in the Amazon regards as irrational the decree limiting deforestation to 20% of a proprietor's land (50% in less-dense forest, insists Mato Grosso). It does not give priority to the most environmentally valuable land. It forces farmers to pay for extra land, which discourages them from protecting it.

The main contender to replace legal reserves is "economic and ecological zoning", which proposes a tactical retreat in the battle against deforestation in order to win it. The government of Parà recently proposed a master plan that blesses deforestation that has occurred so far but raises from 33% to 62% the share of territory off-limits to any but forest-friendly development. For the BR-163, it envisages a strip of development 40km wide on each side of the road, but this would go no farther. The plan accommodates the idea—put forward by social movements—of reserving a mosaic the size of Maine for sustainable use and pure conservation east of the road. This is a surprise coming from a government seen as no friendlier to the forest than Mato Grosso's. Rather than punish people for chopping down trees, the idea is to lure them to "zones of consolidation", where the damage has already been done.

Mind the gap

Visions of the Amazon are converging, yet stop short of consensus. The green-minded federal environment

111

ministry would consider modifying the system of legal reserves, but only after deforestation is brought under control. As for the land along the BR-163, "it is not in the interest of the government to open [new] areas" to development, says João Paulo Capobianco, secretary of forests and biodiversity. But the choice between growth and conservation is a hard one. At the close of an interview, Gabriel Guerreiro, Parà's environment secretary, unexpectedly invites a visitor to tap his teeth: they are false. His teeth fell out when he was an impoverished 15-year-old growing up in the forest. "I don't accept being condemned to poverty," he says.

The path towards a benignly blacktopped BR-163 is as rough as the road itself. Some worry that the quest for governance will be steamrollered in the rush to pave it. Mr Maggi—fearing that sceptics will block the project indefinitely—now lobbies for a southbound rail line as an alternative. The miracle would be a road that promoted Brazil's growth while protecting the indispensable Amazon.

A City of 2 Million Without a Map

Somewhere in this lakeside Central American town, there's a woman who lives beside a yellow car. But it's not her car. It's her address. If you were to write to her, this is where you would send the letter: "From where the Chinese restaurant used to be, two blocks down, half a block toward the lake, next door to the house where the yellow car is parked, Managua, Nicaragua."

Try squeezing that onto the back of a postcard. Come to that, try putting yourself in the place of the letter carriers who have to deliver such unruly epistles. How, for example, would they know where the Chinese restaurant used to be if it isn't there anymore? How would they know which way is "down," considering that "down," as employed by people in these parts, could as easily mean "up"?

How would they know which way the lake lies, when most of the time—in this topsy-turvy capital, punctured by the tall green craters of half a dozen ancient volcanoes—they cannot even see the lake? Finally, how would they know where the yellow car is parked, if its owner happens to be out for a spin?

Somehow, the people who live here have figured these things out. Granted, they've had practice. After all, most Managua street addresses take this cumbersome and inscrutable form. "We don't have a real street map," concedes Manuel Estrada Borge, vice president of the Nicaragua Chamber of Commerce, "so we have an amusing little system that no one from anywhere else can understand."

Welcome to Managua, quite possibly the only place on Earth where upward of 2 million people manage to live, work, and play—not to mention find their way around—in a city where the streets have no names.

No numbers, either. Well, that isn't quite true. A few Managua streets do indeed have conventional names. Some houses even have numbers. But no one hereabouts ever uses them. Why bother? Managuans have their own amusing little system to sort these matters out, a system that has the amusing little side-effect of driving most visitors crazy.

"For people who've just come here," says a long-time Canadian resident of the city, "there's no way on God's Earth that they'd know what you're talking about."

What Managuans are talking about, when all is said and done, is an earthquake that shattered this city three decades ago. Before that time, Managua was an urban conglomeration much like any other, at least in the sense that it had a recognizable center. It also had streets that ran east and west or north and south, and those streets not infrequently bore names. And numbers.

But then, on Dec. 23, 1972, the seismological fault lines that zigzag beneath Managua shifted and buckled, with horrific results. Upward of 20,000 people were killed in the quake, and the city was pretty much reduced to rubble. The catastrophe thoroughly disrupted the old grid pattern of Managua's streets, so the city's surviving residents were obliged to devise a new way of locating things. They started with a landmark—a certain tree, for example, or a pharmacy or a plaza or a soft-drink bottling plant—and they went from there.

Nowadays, for example, if you wished to visit the small Canadian Consulate in Managua, you would present yourself at the following address: *De Los Pipitos, dos cuadras abajo*. In English, this means: From Los Pipitos, two blocks down.

Any self-respecting inhabitant of Managua knows that "Los Pipitos" refers to a child-welfare agency whose headquarters are located a little south of the Tiscapa Lagoon. Managuans also know that *abajo*, in this context, does not mean "down" in a topographical sense. It means "west," because the sun goes down in the west. (By the same token, in Managua street talk, "*arriba*," or "up," means "east." *Al lago*, which literally means "to the lake," is how Managuans say "to the north." For some inexplicable reason, when they want to say "to the south," Managuans say "*al sur*," which means "to the south.")

Just to make a complicated process even more perplexing, Managuans, who normally use the metric system, will often give directions by employing an ancient Spanish unit of measurement called the *vara*. They will say, "From the little tree, two blocks to the south, 50 *varas* to the east." Visitors will therefore need to know how long a *vara* is (0.847 meters). They will also need to know that the "little tree" is no longer little. It is actually quite tall.

A few years ago, the Nicaraguan postal agency considered scrapping the jerry-rigged system of street addresses. But nothing came of the project. Besides, the scheme actually does seem to work. Nedelka Aguilar, for example, has learned that you merely have to have a little faith. Born in Nicaragua, she left as a young girl and spent most of her youth in southern Ontario. Now she lives in Managua once more. Shortly after her return four years ago, she arranged to visit a woman who dwelled at that outlandish address—"From where the Chinese restaurant used to be, two blocks down, half a block toward the lake, next door to the house where the yellow car is parked." By this time, Aguilar spoke the Managua dialect of street addresses well enough to take in the gist of this information. But what about that yellow car?

"I said to the woman, 'How will I find you if the yellow car isn't there?'" Aguilar smiles and shakes her head at the memory. "The woman laughed. She said, 'The yellow car is always there.'"

—Oakland Ross, *The Toronto Star* (liberal), Toronto, Canada, April 21, 2002

AIDS Infects Education Systems in Africa

Yet School Is Critical Factor In Combating Pandemic

Bess Keller

Teachers in Zambia nearly went on strike a few years ago because they weren't paid on time.

Neither of the two government officials responsible for the payroll had reported to work as the salaries came due. One was out sick, almost certainly from an AIDS-related illness. The other was attending to the death rites of someone who had died of the disease.

A strike was averted only when the funeral had taken place and the second official returned to work.

The Rev. Michael J. Kelly, a longtime professor of education at the University of Zambia until his recent retirement, tells this story to bring home the point that the AIDS pandemic raging across sub-Saharan Africa doesn't stop with personal carnage. It also threatens whole systems, including what is arguably the most critical for the region's future—education.

"For countries that are not very rich in managerial personnel, the loss of a few managers can be very deleterious," Father Kelly said.

Similarly, the loss of teachers in the region can sink the quality of education as classes are combined and teachers with fewer qualifications are hired.

"In many African countries, we are talking about educational systems that are already fragile in many ways—they may lack materials; teachers are poorly trained, especially in rural areas," said Cream Wright, the chief of education for UNICEF and a native of Sierra Leone.

In such a situation, any reversal—the death of a teacher, the bereavement of a child, a reduction in budget—can more easily do harm.

The injury goes deeper than it otherwise might because AIDS is destroying families, which undergird the education system. Families are the mainstay of schooling in any country, but in African nations, the family is often the only social safety net that can keep children in school. Now, even that net is seriously frayed by the AIDS-related illnesses and deaths of men and women in their most productive working years.

"Both on the side of schools and of homes, AIDS destroys those coping mechanisms that are in place," Mr. Wright said.

Where rates of HIV infection are high, as they are in much of southern and eastern Africa, experts warn, the effects on social stability and education are so great that young people are being robbed of hope, and national development is being stunted.

And in a final merciless twist, declines in education reduce the chances of arresting the pandemic, since schools may be the best

way to reach uninfected young people with the information, skills, and attitudes that ultimately protect them.

11 Million and Counting

Of the estimated 39 million people worldwide living with the human immunodeficiency virus, for which there is no vaccine and no cure, some 70 percent are in sub-Saharan Africa.

Yet the prevalence of the virus that causes AIDS varies enormously even in this hardest-hit region. In several West African countries, including the most populous, Nigeria, the infection rate is less than 5 percent. But in Botswana and Swaziland in southern Africa, more than 35 percent of the adult population is infected, and the rate continues to rise, according to UNAIDS, the United Nations AIDS coordinating group.

Millions of Africans have died of the disease in the past 20 years. The bereft include 11 million subSaharan children who have lost one or both parents to the disease, making the total number of orphans in the region more than 34 million. The number of AIDS orphans is expected to rise to 20 million by the end of the decade.

The loss of parents affects school enrollment and learning. Families with fewer workers are less likely to be able to afford the costs of school or be able to forgo the labor of a child who is enrolled. Sick relatives make further demands on children, especially girls, who in many African countries devote hours a day to household tasks.

Juliet Chilengi knows firsthand that orphaned children are likely to lose their chance for an education. To save some of them from that fate, Ms. Chilengi founded the New Horizon orphanage in Lusaka, Zambia, where she lives. One teenage girl now with Ms. Chilengi arrived after the aunt with whom she lived judged the girl a handful and "chased her away." Another succumbed to a sexual liaison with an older man because she had no one else to pay for her education, which was cut short anyway when she got pregnant.

Many extended families in her country are breaking under the strain of poverty from unemployment and AIDS, Ms. Chilengi explained.

"There is such a drastic change," agreed Lucy Barimbui, who coordinates anti-AIDS activities for the Kenya National Union of Teachers. "It's no longer the Africa where a child belongs to everyone, and the teachers have to deal with that."

Even where children's material needs are met, grief and insecurity all too easily interfere with learning.

Orphaned by AIDS

Millions of children have been orphaned as a result of the AIDS pandemic in sub-Saharan Africa, and the numbers are expected to climb.

	2001	2010[*]
Angola	104,000	331,000
Benin	34,000	113,000
Botswana	69,000	120,000
Burkina Faso	268,000	415,000
Burundi	237,000	296,000
Cameroon	210,000	677,000
Central African Republic	107,000	165,000
Chad	72,000	132,000
Congo	78,000	112,000
Côte d'Ivoire	420,000	539,000
Djibouti	6,000	15,000
Dem. Rep. of Congo	927,000	1,366,000
Equatorial Gunea	<100	1,000
Eritrea	24,000	55,000
Ethiopia	989,000	2,165,000
Gabon	9,000	14,000
Gambia	5,000	8,000
Ghana	204,000	263,000
Guinea	29,000	57,000
Guinea-Bissau	4,000	13,000
Kenya	892,000	1,541,000
Lesotho	73,000	169,000
Liberia	39,000	121,000
Madagascar	6,000	17,000
Malawi	468,000	741,000
Mali	70,000	117,000
Mozambique	418,000	1,064,000
Namibia	47,000	118,000
Niger	33,000	123,000
Nigeria	995,000	2,638,000
Rwanda	264,000	356,000
Senegal	15,000	23,000
Sierra Leone	42,000	121,000
South Africa	662,000	1,700,000
Sudan	62,000	373,000
Swaziland	35,000	71,000
Tanzania	815,000	1,167,000
Togo	63,000	127,000
Uganda	884,000	605,000
Zambia	572,000	836,000
Zimbabwe	782,000	1,191,000

* Projected SOURCE: UNAIDS and UNICEF

Emory University's public health school in Atlanta and an expert

"In many African countries, we are talking about educational systems that are already fragile in many ways."

Cream Wright

Chief of Education, UNICEF

"We try to talk to them and encourage them, so they don't feel the absence of the parents," said Bartholomew Njogu, the head of Nkubu Primary School in Nkubu, Kenya. But that is a tall order for a school that has no paid counselors and where classes number around 40 children.

Taking Action

Meanwhile, the pandemic has sickened and killed thousands of trained school employees. At one point, experts identified male teachers as being a particularly at-risk population because many are posted to jobs away from their families and have the money for extramarital sexual partners—including readily available female students. In the absence of convincing evidence, however, many experts now believe that teachers are no more likely than others of their area and background to be infected.

Still, classes already swollen by recent guarantees of free primary education in, for instance, Kenya, Malawi, and Zambia, are doubled up when teachers are absent. Rural schools, which are harder to staff, particularly suffer.

To try to preserve its teaching force, Zambia has recently begun offering free antiviral therapy to infected teachers, and a pilot project of several South African teachers' unions will do the same.

Education bureaucracies are generally ill equipped to project manpower needs, and in many countries, internationally imposed fiscal constraints, as well as internal economic ones, have kept hiring of additional school staff members to a minimum. A decline in school quality going back to the 1970s that had begun to be arrested in the 1990s has once again taken hold, many experts say.

In fact, the creeping nature of the crisis erodes improvements just when education needs to be making quantum leaps. Schools should be humane and supportive environments, for instance, free from sexual harassment of girls, who far outpace boys in their rate of HIV infection. Gambia has taken a step in that direction by passing a national law prohibiting male teachers from allowing female students as visitors in their homes. Schools in Zambia, for their part, have been fostering compassion for people with AIDS by distributing AIDS emblems to be worn by all teachers on a certain day of the month. And Kenya has produced a pocket-size AIDS-in-education policy that is to go to all teachers.

On new fronts, some education thinkers are proposing a massive resurrection of boarding schools, which had largely fallen into disfavor with aid donors as expensive artifacts of the colonial era. Though day schools are cheaper, boarding schools might provide orphans with a nurturing community as well as an education. Policy experts also envision more flexible forms of schooling tailored to the needs of youngsters who are caring for family members, bringing in income, or themselves heading families.

"This generation is being forced to take on adult responsibilities earlier than any generation in our historical record," said the Rev. Gary Gunderson, the director of the interfaith health program at on AIDS.

Widespread Pestilence

The rates of people ages 15 to 49 living with HIV or AIDS in 2001 were highest in the southern part of the African continent.

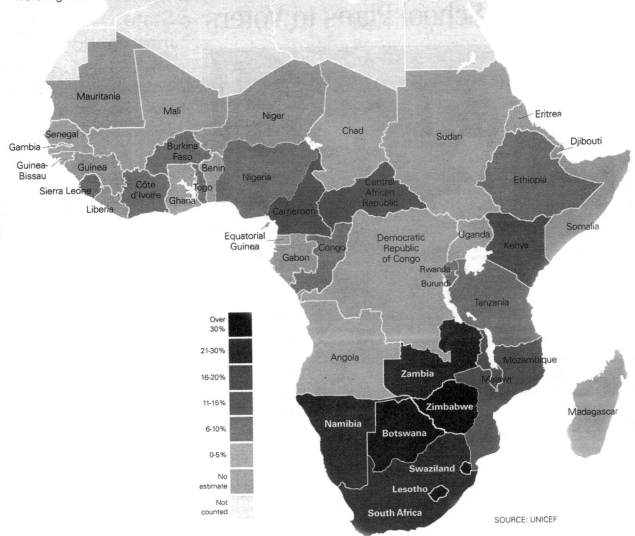

Over 30%

21-30%

16-20%

11-15%

6-10%

0-5%

No estimate

Not counted

SOURCE: UNICEF

Education can literally be a lifesaver for children who must grow up quickly—especially girls, whether by enabling them to learn how to make a living or to protect themselves against the virus, Mr. Gunderson added. "The crisis of HIV/AIDS," he contended, "has way more to do with people in schools than physicians in hospitals."

'Low-Grade Charity'

At the very least, some experts say, schools in sub-Saharan Africa could do a better job of helping children and teenagers avoid the virus. The infection rate among people younger than 15 is low, but then soars, especially for girls. So programs must start early—and go beyond simple awareness. In Cameroon, for example, more than nine in 10 teenagers were aware of AIDS, but fewer than 30 percent knew how to avoid contracting HIV.

Research strongly suggests that the lessons must impart both understanding of the disease and so-called "life skills" for risk management, such as negotiation with a sexual partner and identifying the ways in which a particular social environment poses dangers.

"So far, a lot of resources have been wasted in terms of money, material, and training that don't work," said Mr. Wright of UNICEF. "We're looking to revise [ineffective] programs."

The hurdles to pushing back the pandemic and minimizing its personal and national costs are high but not insurmountable, according to people such as Dr. Peter Piot, the UNAIDS executive director. Anti-AIDS efforts continue to be hampered by prejudice and discomfort because the disease is transmitted mostly by sex and results in death, according to Dr. Piot. He argues that empowerment of people infected by HIV and those most vulnerable to it are among the best antidotes.

No one doubts that millions of Africans have mobilized to try to meet the needs, from compassionate groups operating on members' small contributions to education ministers convening policy and planning groups.

But inadequate funding must be reckoned with. Dr. Piot estimated that the $6.1 billion allocated for response to the pandemic in low- and middle-income countries in 2004 is half of what is needed for the current year. Moreover, the worst-affected African countries, he contended, would gain more from the cancellation of international debts, new trade practices, and better prices for pharmaceuticals than even from increases in rich nations' aid.

"The response already made to AIDS [by Africans] is quite unprecedented," said Mr. Gunderson of Emory. "What has not changed is our response [in the developed world]: We're acting as we've always acted, which is with relatively low-grade charity."

The public-health professor called on educators to add their voices to those of people living with AIDS and their advocates, who played an important role in putting drug treatment within the financial reach of more Africans. Educators in the United States and the rest of the developed world, he said, can grasp the challenges facing African children who have little and have lost much.

"This is an unprecedented crisis, in scale, and nature," Dr. Piot said in a speech in London last month, "and we have no choice but to act in exceptional ways."

UNIT 5

Population, Resources, and Socioeconomic Development

Unit Selections

Key Points to Consider

- How do you feel about the occurrence of starvation in developing world regions?

- What might it be like to migrate from your home to another country?

- In what forms is colonialism present today?

- How can drought impact economic development?

- Why is socioeconomic development in sub-Saharan Africa such a difficult task?

Student Website

www.mhcls.com/online

Internet References

Further information regarding these websites may be found in this book's preface or online.

African Studies WWW (U.Penn)
http://www.sas.upenn.edu/African_Studies/AS.html

Geography and Socioeconomic Development
http://www.ksg.harvard.edu/cid/andes/Documents/Background%20Papers/Geography&Socioeconomic%20Development.pdf

Human Rights and Humanitarian Assistance
http://www.etown.edu/vl/humrts.html

Hypertext and Ethnography
http://www.umanitoba.ca/faculties/arts/anthropology/tutor/aaa_presentation.new.html

Research and Reference (Library of Congress)
http://lcweb.loc.gov/rr/

Space Research Institute
http://arc.iki.rssi.ru/eng/

World Population and Demographic Data
http://geography.about.com/cs/worldpopulation/

The final unit of this anthology includes discussions of several important problems facing humankind. Geographers are keenly aware of regional and global difficulties. It is hoped that their work with researchers from other academic disciplines and representatives of business and government will help bring about solutions to these serious problems.

Probably no single phenomenon has received as much attention in recent years as the so-called population explosion. World population continues to increase at unacceptably high rates. The problem is most severe in the less developed countries where, in some cases, populations are doubling in less than 20 years.

The human population of the world passed the 6 billion mark in 1999. It is anticipated that population increase will continue well into the twenty-first century, despite a slowing in the rate of population growth globally since the 1960s. The first article in this section deals with issues of migration. The next article deals with China's "secret plague:" AIDS. The plight of the Sudanese in the Darfur region follows. The article, "Dry Spell," argues for more proactive governmental responses to drought. The next article reviews programs to turn ocean water into tap water. The next article from *The Economist* deals with approaches to address development problems in Africa. The next article provides a retrospective on the impact of NAFTA in Mexico.

A survey of migration

The longest journey

Freeing migration could enrich humanity even more than freeing trade. But only if the social and political costs are contained, says Frances Cairncross

"**W**ITH two friends I started a journey to Greece, the most horrendous of all journeys. It had all the details of a nightmare: barefoot walking in rough roads, risking death in the dark, police dogs hunting us, drinking water from the rain pools in the road and a rude awakening at gunpoint from the police under a bridge. My parents were terrified and decided that it would be better to pay someone to hide me in the back of a car."

This 16-year-old Albanian high-school drop-out, desperate to leave his impoverished country for the nirvana of clearing tables in an Athens restaurant, might equally well have been a Mexican heading for Texas or an Algerian youngster sneaking into France. He had the misfortune to be born on the wrong side of a line that now divides the world: the line between those whose passports allow them to move and settle reasonably freely across the richer world's borders, and those who can do so only hidden in the back of a truck, and with forged papers.

Tearing down that divide would be one of the fastest ways to boost global economic growth. The gap between labour's rewards in the poor world and the rich, even for something as menial as clearing tables, dwarfs the gap between the prices of traded goods from different parts of the world. The potential gains from liberalising migration therefore dwarf those from removing barriers to world trade. But those gains can be made only at great political cost. Countries rarely welcome strangers into their midst.

Everywhere, international migration has shot up the list of political concerns. The horror of September 11th has toughened America's approach to immigrants, especially students from Muslim countries, and blocked the agreement being negotiated with Mexico. In Europe, the far right has flourished in elections in Austria, Denmark and the Netherlands. In Australia, the plight of the *Tampa* and its human cargo made asylum a top issue last year.

Although many more immigrants arrive legally than hidden in trucks or boats, voters fret that governments have lost control of who enters their country. The result has been a string of measures to try to tighten and enforce immigration rules. But however much governments clamp down, both immigration and immigrants are here to stay. Powerful economic forces are at work. It is impossible to separate the globalisation of trade and capital from the global movement of people. Borders will leak; companies will want to be able to move staff; and liberal democracies will balk at introducing the draconian measures required to make controls truly watertight. If the European Union admits ten new members, it will eventually need to accept not just their goods but their workers too.

Technology also aids migration. The fall in transport costs has made it cheaper to risk a trip, and cheap international telephone calls allow Bulgarians in Spain to tip off their cousins back home that there are fruit-picking jobs available. The United States shares a long border with a developing country; Europe is a bus-ride from the former Soviet block and a boat-ride across the Mediterranean from the world's poorest continent. The rich economies create millions of jobs that the underemployed young in the poor world willingly fill. So demand and supply will constantly conspire to undermine even the most determined restrictions on immigration.

For would-be immigrants, the prize is huge. It may include a life free of danger and an escape from ubiquitous corruption, or the hope of a chance for their children. But mainly it comes in the form of an immense boost to earnings potential. James Smith of Rand, a Californian think-tank, is undertaking a longitudinal survey of recent immigrants to America. Those who get the famous green card, allowing them to work and stay indefinitely, are being asked what they earned before and after. "They gain on average $20,000 a year, or $300,000 over a lifetime in net-present-value terms," he reports. "Not many things you do in your life have such an effect."

Such a prize explains not only why the potential gains from liberalising immigration are so great. It explains, too, why so many people try so hard to come—and why immigration is so difficult to control. The rewards to the successful immigrant are often so large, and the penalties for failure so devastating, that they create a huge temptation to take risks, to bend the rules and to lie. That, inevitably, adds to the hostility felt by many rich-world voters.

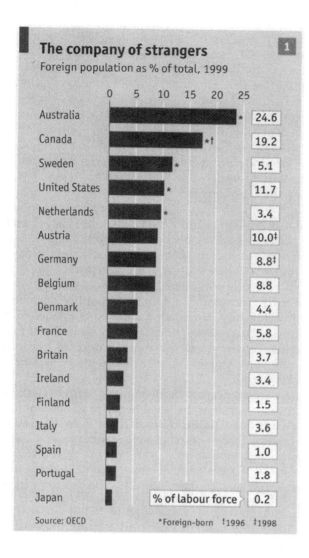

The company of strangers
Foreign population as % of total, 1999

Country	%
Australia	24.6 *
Canada	19.2 *†
Sweden	5.1 *
United States	11.7 *
Netherlands	3.4 *
Austria	10.0‡
Germany	8.8‡
Belgium	8.8
Denmark	4.4
France	5.8
Britain	3.7
Ireland	3.4
Finland	1.5
Italy	3.6
Spain	1.0
Portugal	1.8
Japan	0.2

% of labour force

Source: OECD   *Foreign-born †1996 ‡1998

This hostility is milder in the four countries—the United States, Canada, Australia and New Zealand—that are built on immigration. On the whole, their people accept that a well-managed flow of eager newcomers adds to economic strength and cultural interest. When your ancestors arrived penniless to better themselves, it is hard to object when others want to follow. In Europe and Japan, immigration is new, or feels new, and societies are older and less receptive to change.

Even so, a growing number of European governments now accept that there is an economic case for immigration. This striking change is apparent even in Germany, which has recently been receiving more foreigners, relative to the size of its population, than has America. Last year, a commission headed by a leading politician, Rita Süssmuth, began its report with the revolutionary words: "Germany needs immigrants." Recent legislation based on the report (and hotly attacked by the opposition) streamlines entry procedures.

But there is a gulf between merely accepting the economic case and delighting in the social transformation that immigrants create. Immigrants bring new customs, new foods, new ideas, new ways of doing things. Does that make towns more interesting or more threatening? They enhance baseball and football teams, give a new twang to popular music and open new businesses. Some immigrants transform drifting institutions, as Mexicans have done with American Catholicism, according to Gregory Rodriguez, a Latino journalist in Los Angeles. And some commit disproportionate numbers of crimes.

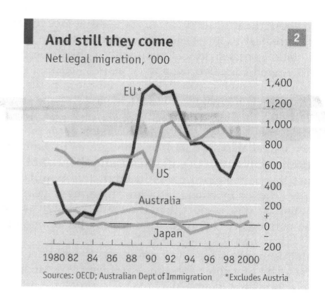

And still they come
Net legal migration, '000

Sources: OECD; Australian Dept of Immigration   *Excludes Austria

They also profoundly test a country's sense of itself, forcing people to define what they value. That is especially true in Europe, where many incomers are Muslims. America's 1.2m–1.5m or so Muslim immigrants tend to be better educated and wealthier than Americans in general. Many are Iranians, who fled extremist Islam. By contrast, some of the children of Germany's Turks, Britain's Pakistanis and France's North Africans seem more attracted to fundamentalism than their parents are. If Muslims take their austere religion seriously, is that deplorable or admirable? If Islam constrains women and attacks homosexuality, what are the boundaries to freedom of speech and religion? Even societies that feel at ease with change will find such questions hard.

No but, maybe yes

Immigration poses two main challenges for the rich world's governments. One is how to manage the inflow of migrants; the other, how to integrate those who are already there.

Whom, for example, to allow in? Already, many governments have realised that the market for top talent is global and competitive. Led by Canada and Australia, they are redesigning migration policies not just to admit, but actively to attract highly skilled immigrants. Germany, for instance, tentatively introduced a green card of its own two years ago for information-technology staff—only to find that a mere 12,000 of the available 20,000 visas were taken up. "Given the higher wages and warmer welcome, no Indians in their right minds would rather go to Germany than to the United States," scoffs Susan Martin, an immigration expert at Georgetown University in Washington, D.C.

Whereas the case for attracting the highly skilled is fast becoming conventional wisdom, a thornier issue is what to do about the unskilled. Because the difference in earnings is greatest in this sector, migration of the unskilled delivers the largest global economic gains. Moreover, wealthy, well-educated, ageing economies create lots of jobs for which their own workers have little appetite. So immigrants tend to cluster at the upper and lower ends of the skill spectrum. Immigrants either have university degrees or no high-school education. Mr Smith's survey makes the point: among immigrants to America, the proportion with a postgraduate education, at 21%, is almost three times as high as in the native population; equally, the proportion with less than nine years of schooling, at 20%, is more than three times as high as that of the native-born (and probably higher still among illegal Mexican immigrants).

All this means that some immigrants do far better than others. The unskilled are the problem. Research by George Borjas, a Harvard University professor whose parents were unskilled Cuban immigrants, has drawn attention to the fact that the unskilled account for a growing proportion of America's foreign-born. (The same is prob-

ably true of Europe's.) Newcomers without high-school education not only drag down the wages of the poorest Americans (some of whom are themselves recent immigrants); their children are also disproportionately likely to fail at school.

These youngsters are there to stay. "The toothpaste is out of the tube," says Mark Krikorian, executive director of the Centre for Immigration Studies, a think-tank in Washington, D.C. And their numbers will grow. Because the rich world's women spurn motherhood, immigrants give birth to many of the rich world's babies. Foreign mothers account for one birth in five in Switzerland and one in eight in Germany and Britain. If these children grow up underprivileged and undereducated, they will create a new underclass that may take many years to emerge from poverty.

For Europe, immigration creates particular problems. Europe needs it even more than the United States because the continent is ageing faster than any other region. Immigration is not a permanent cure (immigrants grow old too), but it will buy time. And migration can "grease the wheels" of Europe's sclerotic labour markets, argues Tito Boeri in a report for the Fondazione Rodolfo Debenedetti, published in July. However, thanks to the generosity of Europe's welfare states, migration is also a sort of tax on immobile labour. And the more immobile Europeans are—the older, the less educated—the more xenophobic they are too.

The barriers need to be dismantled with honesty and care. It is no accident that they began to go up when universal suffrage was introduced. Poor voters know that immigration threatens their living standards. And as long as voters believe that immigration is out of control, they will oppose it. Governments must persuade them that it is being managed in their interests. This survey will suggest some ways in which that might be done.

Irresistible attraction

Who moves, and why

LEAVING one's home to settle in a foreign land requires courage or desperation. No wonder only a tiny fraction of humanity does so. Most migration takes place within countries, not between them, part of the great procession of people from country to town and from agriculture to industry. International migrants, defined as people who have lived outside their homeland for a year or more, account for under 3% of the world's population: a total, in 2000, of maybe 150m people, or rather less than the population of Brazil. Many more people—a much faster-growing group—move temporarily: to study, as tourists, or to work abroad under some special scheme for a while. However, the 1990s saw rapid growth in immigration almost everywhere, and because population growth is

slowing sharply in many countries, immigrants and their children account for a rising share of it.

Counting migrants is horrendously difficult, even when they are legal. Definitions vary. Some countries keep population registers, others do not. The visitor who comes for a holiday may stay (legally or illegally) to work. Counting those who come is hard, and only Australia and New Zealand rigorously try to count those who leave. So nobody knows whether the rejected asylum-seeker or the illegal who has been told to leave has gone or stayed. But the overall picture is one of continuing growth in the late 1990s.

Between 1989 and 1998, gross flows of immigrants into America and into Europe (from outside the EU) were similar, relative to population size. About 1m people a year enter America legally, and some 500,000 illegally; about 1.2m a year enter the EU legally, and perhaps 500,000 illegally. In both America and Europe, immigration has become the main driver of population growth. In some places, the effects are dramatic. Some 36% of New York's present population is foreign-born, says Andrew Beveridge, a sociology professor at Queens College, New York: "It hasn't been that high since 1910," the last peak.

America at least thinks of itself as an immigrant land. But for many European countries the surge of arrivals in the 1990s came as a shock. For example, the Greek census of 2001 found that, of the 1m rise in the population in the previous decade (to 11m), only 40,000 was due to natural increase. "In a decade, Greece has jumped from being one of the world's least immigrant-dense countries to being nearly as immigrant-dense as the United States," notes Demetrios Papademetriou, co-director of the newly created Migration Policy Institute in Washington, D.C.

Asia too saw a burst of immigration in the 1990s, propelled initially by the region's economic boom. Foreign workers accounted for an increasing share of the growth in the labour supply in the decade to the mid-1990s. Chris Manning, an economist at the Australian National University in Canberra, reckons that foreign workers made up more than half the growth in the less-skilled labour force in Malaysia, perhaps one-third of the growth in Thailand and 15–20% of the growth in Japan, South Korea and Taiwan.

What makes all these people move? In the past, governments often imported them. In Europe, migration in the 1950s and 1960s was by invitation: Britain's West Indians and Asians, for example, first came at the government's request. Britain's worst racial problems descend from the planned import of textile and industrial workers to northern England. Now the market lures the incomers, which may produce less disastrous results.

Three forces often combine to drive people abroad. The most powerful is the hope of economic gain. Alone, though, that may not be enough: a failing state, as in Somalia, Sri Lanka, Iraq or Afghanistan, also creates a powerful incentive to leave. Lastly, a network of friends and relatives lowers the barriers to migrating. Britain has many Bangladeshi immigrants, but most come from the single rural district of Sylhet. Many host countries "specialise" in importing people from particular areas: in Portugal, Brazilians account for 11% of foreigners settling there; in France, Moroccans and Algerians together make up 30% of incomers; and in Canada, the Chinese share of immigrants is more than 15%.

Most very poor countries send few people abroad. Immigration seems to start in earnest with the onset of industrialisation. It costs money to travel, and factory jobs provide it. That pattern emerges strikingly from a study by Frank Pieke of the Oxford University of emigration from China's Fujian province. He describes how internal and overseas migration are intertwined. Typically, a woman from a family will go to work in a factory in a nearby province, supporting a man who then goes abroad and probably needs a few months to find himself a job.

Incentives to go, incentives to stay

Net immigration flows continue as long as there is a wide gap in income per head between sending and receiving countries. Calculations by the OECD for 1997 looked at GDP per head, adjusted for purchasing power, in the countries that sent immigrants to its rich members, and compared that figure with GDP per head in the host country. In all but one of its seven largest members, average annual income per person in the sending countries was less than half that of the host country.

Migrant flows peter out as incomes in sending and host country converge. Philip Martin, an economist from the University of California at Davis, talks of a "migration hump": emigration first rises in line with GDP per head and then begins to fall. Migration patterns in southern Europe in the 1980s suggested that the turning point at that time came at just under $4,000 a head. In a study for the European Commission last year of the prospective labour-market effect of EU enlargement, Herbert Brücker, of Berlin's German Institute for Economic Research (DIW), estimated that initially 335,000 people from the new members might move west each year, but that after ten years the flow would drop below 150,000 as incomes converged and the most footloose had gone. Net labour migration usually ends long before wages equalise in sending and host countries.

Migrants do not necessarily come to stay. They may want to work or study for a few months or years and then go home. But perversely, they are more likely to remain if they think that it will be hard to get back once they have left. "If you are very strict, you have more illegals," observes Germany's Ms. Süssmuth.

There has always been a return flow of migrants, even when going home meant a perilous return crossing of the Atlantic. According to Dan Griswold of the Cato Institute, a right-of-centre American think-tank, even in the first decade of the 20th century 20–30% of migrants eventually went home. And where migrants are free to come and go, many do not come in the first place. There is no significant net migration between the United States and Puerto Rico, despite free movement of labour. "It's expensive to be underemployed in America," explains Mr. Griswold. But in Europe, with its safety net of welfare benefits, the incentives to have a go are greater.

Tougher border controls deter immigrants from returning home. A book co-authored by Douglas Massey of the University of Pennsylvania, "Beyond Smoke and Mirrors", describes how in the early 1960s the end of a programme to allow Mexicans to work temporarily in America led to a sharp rise in illegal immigrants. Another recent study, published in *Population and Development Re-*

The going rate
Payments to traffickers for selected migration routes, $ per person

Kurdistan-Germany	3,000	North Africa-Spain	2,000-3,500
China-Europe	10,000-15,000	Iraq-Europe	4,100-5,000
China-New York	35,000	Middle East-US	1,000-15,000
Pakistan/India-US	25,000	Mexico-Los Angeles	200-400
Arab states-UAE	2,000-3,000	Philippines-Malaysia/Indonesia	3,500

Source: "Migrant Trafficking and Human Smuggling in Europe", International Organisation for Migration, 2000

view, also links tighter enforcement to a switch from temporary to permanent migration. Its author, Wayne Cornelius, says the fees paid to *coyotes*, people who smuggle migrants, have risen sharply. He found that, when the median cost of a *coyote*'s services was $237, 50% of male Mexican migrants went home after two years in the United States; but when it had risen to $711, only 38% went back. And, whereas the cost of getting in has risen (as have the numbers who die in the attempt), the cost of staying put has declined, because workplace inspections to catch illegals have almost ceased. The chance of being caught once in the country is a mere 1–2% a year, Mr. Cornelius reckons. So "The current strategy of border enforcement is keeping more unauthorised migrants *in* the US than it is keeping *out*."

Tighter controls in Europe are probably creating similar incentives to stay rather than to commute or return. A complex, bureaucratic system designed to keep many willing workers away from eager employers is bound to breed corruption and distortion. And the way that rich countries select immigrants makes matters worse.

CHINA'S SECRET PLAGUE

How one U.S. scientist is struggling to help the government face up to an exploding AIDS crisis

BY **ALICE PARK**
Kunming

THEY LINE THE DUSTY ROADS OUTSIDE THE tiny villages of China's Henan province, several hours' drive from Beijing—mounds of dirt funneled into crudely shaped cones, like a phalanx of earthen bamboo hats. To the uninitiated, they look like a clever new way of turning over fields—an agricultural innovation, perhaps, meant to increase crop yields. But the locals know the truth. Buried under the pyramids, which now number in the thousands, are their mothers and fathers, brothers, sisters and cousins, all victims of AIDS. Like silent sentries, the dirt graves are a testament to China's worst-kept secret.

They are the reason Dr. David Ho has come to China. The New York City-based virologist was named TIME's 1996 Person of the Year for his pioneering work on the drug therapies that have largely quelled the AIDS epidemic in the U.S. and Europe. Now Ho is confronting the AIDS virus in its most populous stronghold. Up to 1 million Chinese are HIV positive, and that number could easily grow to 10 million by 2010, according to the Joint U.N. Program on AIDS. If current trends continue for another decade or so, China could overtake Africa, where 29 million people have been infected with the virus.

It's to head off that scenario that Ho has traveled more than a dozen times to China over the past three years, setting up labs, visiting clinics, gathering blood samples, educating health workers and negotiating the intricately layered bureaucracy of the Chinese health establishment. Ho's efforts—and those of other AIDS activists—finally paid off last week when, on World AIDS Day, the Chinese government took a lesson from its sluggish response to the severe acute respiratory syndrome (SARS) epidemic and launched its first big AIDS public-awareness campaign, complete with posters, TV spots and an unprecedented visit by Premier Wen Jiabao to a Beijing hospital, where he shook hands with AIDS patients.

TIME accompanied Ho and his team from the Aaron Diamond AIDS Research Center (ADARC) for two weeks earlier this year as he traveled from Kunming, the cosmopolitan capital of Yunnan province, where his drug-treatment and vaccine projects are based, to the remote border town of Ruili, where heavy heroin trafficking and a thriving sex trade create a perfect HIV breeding ground, to Beijing, for his meetings with party leaders, including the newly appointed Minister of Health, Wu Yi. Everywhere Ho went, his mission was the same: to persuade Chinese officials to step up their modest anti-AIDS efforts and commit the resources necessary to launch a comprehensive nationwide program, modeled on the projects he has begun in Yunnan.

Kunming, Yunnan

THE NEATLY DRESSED HUSBAND AND WIFE ARE IN THEIR 50S and comfortably average looking. Their once-smooth dark skin is now veined and burnished to a proud sheen, reflecting the decades of hard work they have put into raising a family, earning their salaries and, now, battling HIV.

They seem out of place in the world of AIDS. Neither injects drugs. Neither has had any contact with the sex trade. But they represent the newest and most troubling front in China's war against the AIDS virus. As in other countries hit by HIV, the epidemic in China began in the margins of society—among migrant workers, drug users and prostitutes—and then gradually entered the mainstream population. In China this process was facilitated by the government, which, through the tragic mismanagement of its blood-buying program in the early 1990s, permitted blood-collecting practices that ended up contaminating the country's blood supply with HIV. Anyone who gave blood or received a transfusion during that period was at high risk of contracting the virus—and then passing it on to his or her partners during intercourse.

That was how this couple, who declined to give their names, got the AIDS virus. They have kept it a secret from everyone but their immediate family, preferring not to risk being ostracized by their community. "Nobody knows," says the wife quietly. "They would not understand." The husband, as far as they can determine, was the first to get infected, perhaps from blood transfusions during surgery. It wasn't until his wife required an operation in 2001, however, that they were both found to be HIV positive. "I could not believe it," she says. "I told them they were totally wrong, that their detection was wrong. I heard reports that there was HIV in China, but that was mainly from people who traveled overseas. We never thought the virus would get here, in our family."

In a way, they are the lucky ones. Along with 68 other patients, they are part of a treatment program that Ho established in Kunming. There they will get the latest antiretroviral medications and the same careful monitoring that AIDS patients in the U.S. receive, including regular measurements of their viral loads and their immune-cell counts and tests to determine how quickly the virus is mutating to resist the drugs.

The epidemic began in the margins of society—among migrant workers, drug users and prostitutes

The vast majority of the Chinese who are HIV positive have no such access and must make do with drugs that treat the side effects of the disease—antibiotics for mouth sores and pneumonia, creams for skin lesions. Others rely heavily on traditional Chinese herbal medicines, which have no documented record of success. And even for those who are able to squeeze into one of the small studies supported by foreign aid groups, there is no guarantee of receiving proper follow-up care. "We have heard of places in China where the drugs are delivered but there is no training of the doctors in how to use them," says Ho. "We stress to them that drug treatment for AIDS is not like food relief, where the food is just dropped off."

As powerful as the AIDS drugs are, HIV mutates so rapidly that if the antiretroviral compounds are not properly administered, they are quickly rendered useless not just for that patient but for every other patient exposed to the mutated virus. It's a concept that is difficult for even the best-intentioned patients here to appreciate. TIME spoke with a patient advocate, 31, who goes by the pseudonym Ke'Er. He was infected after selling blood and was admitted to a study in Beijing that provided free U.S. antiretroviral drugs, but he accidentally left his two-month supply on the train after his most recent visit to the city. "I dared not tell my doctor," he said, "because I felt bad that I was offered this opportunity but I lost my medicine. So I found a Thai drug cocktail that is similar, and I'm taking that now." He doesn't know what the Thai drugs are but was assured by a doctor in his village that they would help. Chances are they won't.

Even the best AIDS drugs properly administered can do only so much. What doctors really need to head off a runaway epidemic is an effective vaccine. In fact, it was a vaccine trial that took Ho to China in the first place. In a way, China is an ideal place to conduct vaccine research. Because it is home to huge numbers of people who are HIV negative but at high risk of developing AIDS, Ho will be able to inoculate some of them with his vaccine and find out whether they can generate an immune response robust enough to protect them in case of a future exposure to HIV.

He is scheduled to inoculate his first healthy volunteers in New York with the U.S. version of the vaccine this week. Before he can begin testing a vaccine in China, however, Ho needs to know more about the virus strains circulating there. To protect against HIV, any experimental AIDS vaccine must be designed to match the rapidly changing strains moving through a population. Ho needs access to the blood of a lot of HIV-positive patients, so when he started looking for a place in China to conduct his trials, he turned first to Yunnan, a province with one of the greatest numbers of HIV and AIDS cases. His hope was that health officials there, who see the daily toll the disease takes, would be more willing to accept help from an outsider. It wasn't that simple.

Tracking HIV

YUNNAN IS CHINA'S FOURTH LARGEST province and historically one of its most mysterious and remote. (Its picturesque landscape of verdant hills and rustic villages inspired the legend of Shangri-La.) Its distance from the political leaders in Beijing has traditionally made it something of an outlaw province, home to dozens of minority groups and, in centuries past, feudal warlords who ruled with nearly absolute control. Today it is the gateway for heroin traffic that drifts into China from Burma, Vietnam and Laos.

As many as 10 million Chinese could be infected by 2010

Scattered along the drug route is China's largest concentration of heroin addicts. Yunnan has the highest IV-drug-use rates in China, and a recent U.N. AIDS report estimates that anywhere from 50% to 80% of the users are carrying the AIDS virus. HIV spread via unprotected sex is also on the rise here, accounting for 15% of HIV infections in 2000. All told, say health officials in Yunnan, this single province accounts for one-third of China's reported AIDS cases.

Given the Chinese penchant for careful record keeping, it's no surprise that officials here have been collecting and analyzing information on these cases for more than a decade. But access to the data—especially for outsiders—has been carefully guarded. The man in charge of generating the statistics is Dr. Lu Lin, director of the Yunnan Center for Disease Control (CDC), who has been monitoring HIV infection among the highest-risk groups in nearly 50 sites around the province since 1991.

A former prison guard with hooded eyes and a buzz cut who, at over 6 ft. tall, towers over most Chinese, Lu might seem a tough nut to crack. But when Ho approached Lu and his colleagues three years ago with a proposal to collaborate on vaccine trials, Ho was surprised by the response he got. They were eager to cooperate, he recalls, but had little interest in a vaccine. They were more concerned with helping those already struggling with the disease. "We wanted to push the vaccine," says Ho, "and they wanted to get more treatment for patients, more trained people and better labs to take care of the patients."

So for the past two years, Ho has retreated from his vaccine agenda and set up the pilot drug-treatment program in Kunming. Using funding from both ADARC and private donors, he has also built a clinic, set up a virology lab capable of performing basic viral-load tests and put together a state-of-the-art immunology lab—all of which will eventually absorb the testing required for the future vaccine studies.

In return, Ho has asked for access to the blood samples—some 24,000—collected from HIV patients throughout the province over the years. The samples will give him critical information about which populations in Yunnan would be suitable as the first subjects for his vaccine trials. "We realized we needed a quid pro quo," he says.

As part of that exchange, Lu's CDC team shared with Ho, in the first presentation of its kind to anyone outside the Chinese government, the details of AIDS penetration in Yunnan. Last March Lu informed Ho that in a 2002 survey of high-risk populations, 43% of IV drug users had shared needles with others in the past month, and that among female sex workers, 89% were unaware of their risk of contracting HIV. A majority of sex workers, about 60%, reported inconsistent condom use. Since they have begun collecting data, says Lu, there has been a 25% to 30% increase in HIV cases among IV drug users in the province. The incidence of HIV infection among sex workers has also risen steadily.

It was what Ho suspected but could never confirm without the data. Clearly, the few programs that the Chinese had put in place—distributing condoms and educating people about the dangers of unprotected sex—were having little effect on the spread of HIV, and most of the population was still both misinformed and uninformed about how dangerous the virus is. "We all appreciate that the epidemic in China was bigger than our expectations," he says. "We found ourselves taking on issues beyond just our research agenda. We realized that with a few more partners, we could—and should—do more educating, treating and training of people about AIDS in China."

To broaden the scope of his efforts, Ho enlisted the support of the newly appointed director of the province's Bureau of Health, Chen Juemin. Chen, to Ho's relief, is intent on addressing the AIDS epidemic in his province and is eager to have Yunnan serve as a testing ground for programs that Minister of Health Wu in Beijing will consider for the rest of the country. "This situation will not just go away," Chen told TIME. "We probably lost a chance [of controlling AIDS] because we did not open up publicly about our HIV work in the early 1980s. We didn't realize then that the disease was so serious and could spread so fast."

Mangxi, Yunnan

THE LAB, SUCH AS IT IS, CONSISTS OF just three rooms squeezed into a four-story building deep in Yunnan's southwestern town of Mangxi. The building has no elevator, and the external stairwell is bathed in the steamy heat that washes the entire region. Inside, however, in stark contrast to its tropical-outpost surroundings, are a few jewels of the modern microbiology trade—a state-of-the-art freezer for storing blood samples and an enzyme-linked immunosorbent assay (ELISA), a machine for screening HIV that can identify specific antibodies to the virus.

The equipment, including a computer and fax machine, all donated by Ho, will enable Mangxi to share vital data with Kunming, 280 miles away, and with Ho's group in the U.S. Yunnan's first case of HIV infection was discovered in Mangxi in 1989. Presumably the virus has been circulating here the longest; being able to include patients from the region in his study will enable Ho to tell how quickly the virus is mutating and which strains should be part of his experimental inoculation.

In Mangxi, Ho's priority is to sign up subjects, not an easy task when many of the prospective candidates are IV drug users and live in remote, largely inaccessible villages without telephones or newspapers; in fact, few of them can even read. Local health officials conduct their prevention efforts the old-fashioned way—going family to family, teaching couples how to use condoms and warning the young about the dangers of sharing needles.

One likely source of research subjects is the drug-rehabilitation camps that are blossoming all over Yunnan. A drug user picked up by the police is often forced to serve a mandatory three-month sentence in a rehabilitation camp, where calisthenics, lectures and daily treatment with a Chinese version of methadone are supposed to curb the addict's habit. Up to 20% of the inmates, by the guards' rough estimates, are HIV positive; because they are registered by the police, they can be tracked after they leave the camps. Eventually Ho wants to find and monitor 500 HIV-negative patients in the Mangxi area who are at high risk of becoming infected. Merging information on how many in this population eventually become HIV positive with data from the urban residents of Kunming will help him measure how quickly the virus is spreading.

Henan province

CHUNG TO, FOUNDER AND DIRECTOR of the nonprofit Chi Heng Foundation in Hong Kong, is one of the few outsiders who has penetrated the state-imposed isolation of the so-called AIDS villages in central China. He is all too familiar with the plight of small children orphaned by the disease. On a recent visit to a village in Henan, he watched an 8-year-old boy taking his father out for a walk. The boy was pushing his father along in a creaky wooden cart. The man was dying of AIDS and had been confined to his bed for weeks, too weak to walk. His son suggested the cart, hoping that a little fresh air would energize his ailing parent. A few weeks later, the father was dead.

"It was an unforgettable scene," says To. Using his own funds and donations, To has been helping these children continue their schooling, giving them a chance to free themselves from the taint of having a parent—or both parents, in some cases—die of AIDS.

In heavily affected provinces like Henan, Hebei and Shaanxi, an entire generation is vanishing in the shadow of AIDS. In family after family, mothers and fathers are dying, leaving as many as 200,000 children in Henan alone either parentless or in the care of aging grandparents. Ho and his colleagues were the first foreign group officially allowed to visit one of its villages, Wenlou. At the local hospital, only two doctors care for more than 1,000 HIV-positive patients, and they were trained not by the Chinese health system but by one of Ho's colleagues based in China.

Here, unlike in Yunnan, HIV is spread not through illegal behavior but through blood donation. In the early 1990s, the Chinese leadership launched a blood drive and paid donors for their plasma. It was a program intended to benefit all Chinese—the poor by giving them a way to supplement their income, and the rest of China by replenishing the national blood banks' dangerously low stocks. "It was like a poverty-relief program," says a Henan resident who gave plasma in 1993 and became infected. Through campaigns in the villages and schools, the government encouraged rural farmers and factory workers to sell their plasma for 40 yuan ($5). The good intentions backfired when "bloodheads," as some of the unofficial blood collectors came to be known, found a way to extract more plasma from fewer donors. Those running some stations pooled and processed the blood. Then they sent the plasma, containing useful proteins, to the blood banks and reinjected red and white blood cells, which can house HIV, into the donors. This enabled people to give several times a day, and nobody seemed to realize how dangerous the practice was. Infected blood now flowed through hundreds of thousands of residents in the central provinces, shifting the epicenter of AIDS cases, many experts believe, from Yunnan to the heart of China.

Henan and its neighbors, Ho has decided, cannot wait for his program to become established in Yunnan. In his proposal to the Ministry of Health, Ho has modified his plan to include testing, treatment and prevention projects for Henan and Yunnan. "They desperately want help," he says of the doctors he met in Wenlou. "They obviously have the data on AIDS patients but are afraid to show us."

Even today, 1 out of 5 Chinese have never heard of AIDS

That fear is well founded. Adding to the stigma surrounding AIDS in these villages is the role that local leaders played in the blood-buying program. "Many government officials made a lot of money," says the patient advocate who calls himself Ke'Er. To protect themselves, they wrapped their villages in the cloak of state secrecy, effectively sealing off AIDS patients from foreign aid groups as well as health officials from other provinces. AIDS-care centers still won't put the word AIDS on their doors, opting instead for such intentionally obscure labels as "home garden."

To break through this barrier of fear, Ho has encouraged Health Minister Wu to visit the AIDS villages in Henan. Wu's visit would be the first by someone in her post and would send, Ho hopes, a powerful message that the government is more interested in controlling the epidemic than in assigning blame. Wu was appointed Health Minister when her predecessor, with whom Ho had begun his project, was fired by the Communist Party for mishandling the SARS outbreak—denying its existence until the epidemic was out of hand. "SARS was a big kick in the pants for China," Ho says. "They were tainted by the SARS experience, and the health officials there now want to do the right things with AIDS."

Ho doesn't expect miracles. Many of the cultural traditions that make it difficult for the Chinese people and their government to openly address a sexually transmitted disease are too deeply rooted for one man to change. A recent survey by Futures Group Europe and Horizon Research Group revealed that 20% of Chinese still have not heard of AIDS and that only 5% have had an HIV test. Ho is convinced that even if just part of his program is put in place, it will save lives. "If we had known how difficult the process was going to be, I'm not sure we would have embarked on it," he says, reflecting on his work of the past three years. "We put up with a lot. But as AIDS researchers, we could not continue to be distant from the vast majority of patients."

The work, after all, is just beginning. Ho's team in New York City has analyzed the first material from the blood samples. "It looks really good," says Ho, visibly brightening at the prospect of finally starting up his vaccine studies. "Any one of the sites in Yunnan would work well for a vaccine trial." Starting those trials will mean China is that much closer to controlling HIV and slowing the spread of those earthen graves of family members claimed by AIDS.

Farms Destroyed, Stricken Sudan Faces Food Crisis

Two Years of Violence Leaves Growers Afraid to Plant; Officials Downplay Woes

Cutting Down Mango Trees

Roger Thurow

FUR BARANGA, Sudan—After the killings, the rapes, and the expulsion of nearly two million farmers from their land, the people of Darfur are now facing a new threat—the worst food shortage in decades.

"The harvest was so bad, there is no Darfur sorghum," says Khaltom Khalid, a trader in the local market. Her meager offerings bear witness to the consequences of the intentional destruction of farms across this vast region. All she has for sale is a small mound of the grain, carried by donkeys from the neighboring country of Chad. The price has doubled from this time last year and she offers a measuring cup half the normal size to scoop up smaller portions. But she has so few customers that she spends much of the day sleeping on a reed mat in the sand beside her sorghum, which is used in better times to make flat bread and porridge.

"Prices are high, nobody can afford to buy," she says. "Everybody is getting hungry."

For two years, marauding militias composed mainly of Arab nomads and cattle herders have attacked Darfur's African farmers in a battle over arable land. United Nations agencies estimate more than 70,000 people have already died. Now the food crisis is giving the conflict a deadly

new momentum—threatening both the farmers and those who brutalized them.

Averting a hunger disaster here is being made more difficult by at least two factors: The world's humanitarian attention has turned to focusing on victims of the tsunami in Asia. And since hundreds of farming villages have been flattened and agricultural equipment and seed stock destroyed, any recovery will take much longer.

Last week, a U.N. commission probing the Darfur violence said Sudanese government forces and allied militias have committed atrocities on a "widespread and systematic basis." It recommended these "serious violations of international human rights" be referred for prosecution in the International Criminal Court. Sudan's government denies participation in the brutality, saying it is waging a counterinsurgency campaign.

The current scarcity of food, and the harsh market forces it has unleashed, have become the new agents of the violence that has been labeled "genocide" by the U.S. With surviving farmers huddling in domestic refugee camps, two harvests have already been lost. And a third ruinous year looms, as farmers too afraid to leave the camps are giving up on this spring's planting season.

"All the indicators are there for a famine," says Marc Bellemans, the Sudan emergency coordinator for the U.N.'s Food and Agriculture Organization. In a report to fellow U.N. agencies late last year, the FAO warned "a humanitarian crisis of unseen proportions is unfolding in the Darfur region, with conditions similar to those preceding the 1984 famine." That famine, ignited by drought, killed an estimated 100,000 in Darfur and more than one million throughout northeastern Africa.

Government agriculture officials in West Darfur, the most fertile region in the area, say last year's harvest yielded about 48,000 tons of grains—less than a fifth of the amount needed for the region to feed itself. Prices of everything from sorghum to peanuts have doubled or tripled across Darfur, putting much of the food beyond the reach of the impoverished population.

"We used to grow everything we needed for ourselves, and the surplus we sold in the market. Peanuts, tomatoes, okra, sesame, wheat, sorghum," says Khamis Adam Hassen Okey, the leader of Andarbrow, a farming village in West Darfur. It was leveled during an attack in October 2003, he says, just as the harvest began.

Mr. Okey says 46 villagers were killed and five women were raped. The survivors among the village's 150 families fled to Fur Baranga, where they now live in small thatch huts, in the courtyard of an unfinished hospital and are cared for by international relief agencies. "We are farmers," he says. "But if we go out from this place to plant, we will be killed."

Even Arab camel and cattle herders—many of whom have taken part in attacking the farmers and now graze their animals on land where crops once grew—complain of not having enough to eat. In some parts of Darfur, the fighting has blocked herders from moving their livestock north to the markets in Libya and Egypt. With their sales down, they don't have enough money to pay the higher prices for what little food is available.

More to Feed

So far, full-blown famine has been kept at bay with help from food rushed into Darfur by the U.N.'s World Food Program. The U.S. is the program's largest donor. In December, the WFP fed about 1.5 million people in Darfur. But the agency is predicting a steep rise in the number it will need to feed this year—up to at least 2.7 million a month—as displaced farmers continue to file into camps and city residents and herders become more desperate. The amount of food WFP says it needs to bring to Darfur this year is triple what it needed in 2004, or more than 450,000 tons.

Jan Pronk, the U.N's chief envoy in Sudan, says the cumulative effect of one failed harvest after another, along with rising prices and malnutrition rates, could eventually leave all six million Darfurians in need of food aid. Surveying the economics now in play, he says, "the future in 2005 is bad."

In famines where drought has killed crops, farmers are largely able to recover when the rains return. Darfurians have had experience with this; they protect seeds for the next planting season and try to keep their animals alive. This time, the hunger has been willfully engineered by destroying all aspects of the agricultural system. Seed stocks have been burned, animals stolen or killed, and the tools of cultivation, such as hoes and tractors, smashed.

While the 2003 famine in Ethiopia was one of the worst for the number of people made vulnerable—12 million were kept alive by food aid—cultivation picked up again the following year. The current Darfur crisis, however, will likely continue even after the farmers return to their land, because the very means of their livelihood have been destroyed.

As the need for longer-term aid in Darfur escalates, the world's attention has shifted to the tsunami devastation in Asia. Last year, the U.N. branded Darfur the "world's worst humanitarian crisis." Post-tsunami, Ramiro Lopes da Silva, the WFP's Sudan director, wryly refers to Darfur as "what is now the number two emergency in the world."

In December, a few weeks before the tsunami, Carlos Veloso, the WFP's emergency coordinator for Darfur, appealed to donor countries to speed up contributions. He said more than half of the $438 million of food aid needed in Darfur for 2005 must be delivered by the end of January, to insure food would be in place before rains make overland transport nearly impossible and isolate tens of thousands of people. Mr. Veloso said that in addition to food donations—wheat, beans, cooking oil and a corn-soya blend make up the standard ration—cash contributions were needed to buy dozens of heavy-duty trucks to haul the food across the desert.

The U.S. responded with a donation of 200,000 tons of wheat, valued at $172 million. Then the tsunami hit. Donations of other food commodities have been slow to arrive; so has money for the trucks. "The window of opportunity is narrowing," says Mr. Veloso. If enough food isn't available for Darfur, the WFP may be forced to reduce the size of the monthly rations, or limit the number of recipients.

Even when Darfur dominated the humanitarian spotlight last year—nearly 100 relief agencies flocked in to help with water, sanitation and health care—Darfur's farming needs were overlooked. The U.N. says the agriculture section of its appeal, which would have provided seeds and tools to help farmers, received less than one-fifth of funding requirements.

For years, the farmers and herders of Darfur, though wary of each other, had a system for sharing the land and providing food to each other. The war between the largely Arab government and rebel groups seeking a greater voice in politics and economic development has ruined that. Now, competition for scarce food, blocked nomadic routes and the scramble to get international food aid is only heightening longstanding mistrust between the herders and farmers. "Conflict leads to scarcity and scarcity leads to more conflict," says Mr. Pronk.

Although the U.S. has labeled the events in Darfur a genocide and other countries have called it an ethnic cleansing, no western military force has joined a peacekeeping mission that now consists of a small African Union deployment. U.S. diplomats complain that their attempts to get the U.N. to impose sanctions on Sudan's government have been stymied by other countries, notably China, which is a major partner in Sudan's oil production.

Government officials in Khartoum downplay the food scarcities, and have resisted pressure from the U.N. to stabilize prices by shifting 100,000 tons of food from other regions to Darfur. "There is a food gap in Darfur, but it's not so significant," says Ahmad Ali El Hassan, the director of rain-fed agriculture. "Humanitarian assistance will fill the gap."

He insists Darfur's farmers, despite continuing security threats, will leave the camps and return to their farms to plant this spring. He concedes the farmers' seeds, tools and livestock have been destroyed in the war, which the government blames on tribal conflicts. Still, he says, "God willing, we'll get a good crop."

But farmers in the refugee camps say they have given up hope of returning in time to plant, fearing attacks from the same militias—known as the Janjaweed—that drove them away in the first place. "No way I'm going back this year," says Matair Abdall, emphatically shaking her head.

Her village of Willo, she says, was destroyed by the Janjaweed, who burned the fields, knocked down huts and chased away farmers in late 2003. Ms. Abdall says she, her husband and four children walked three hours to a refugee camp. Last spring, she returned to plant sorghum. As the crop started growing, cattle ate some of it, she says, but the herders who have taken over her village let the rest grow. At harvest time, she filled six 200-pound bags, which she says would have fed her family for much of the year.

Men With Guns

"Then five men with guns on horses, they were the Janjaweed, surrounded me and said, 'Give us the harvest.' I was afraid and gave it to them," says Ms. Abdall. "I won't go through that again this year."

Instead, she sits in one of the camp's activities center with two dozen women, weaving baskets to earn some money.

"We'll never feel as safe farming again," says Asha Ashagg, 35, who nurses the youngest of her five children while she weaves. She says she was driven out of her village by the Janjaweed. Once she returned home to check on her garden and fruit trees, she says, and found that even her mango trees and banana plants had been cut down. "They do it so we won't return," she says. When she was at her village, she says, "the Janjaweed came and asked, 'Why have you come back? What is your tribe?'." She says she ran back to the camp and hasn't returned home since.

The road from her refugee camp to Fur Baranga, a 100-mile dirt path through scrub brush, is lined with destroyed villages. Scattered clay water pots, the remains of small mud-brick houses and abandoned schools and health centers are evidence of people uprooted. Cows, sheep, goats and camels now graze where crops once grew.

Mr. Okey's home of Andarbrow is one of the empty roadside villages. "Only after all the guns are collected will we go home," he says. In the meantime, his villagers, crowded into the Fur Baranga camp, rely on food from the World Food Program, and the little money they make cleaning up after traders in the market.

They also receive handouts from residents of Fur Baranga, who have been sharing with fellow tribal members. But now the cupboards of the residents are getting bare, too, and high prices prevent restocking. "Even the price of dried okra is too much," says Yahya Arabi Yahya, complaining of paying double the price for a vegetable that once grew in abundance. "And now water is a problem," he says, noting that water prices have risen as the influx of displaced farmers has increased demand.

Many of the African residents of Fur Baranga and other towns swamped by displaced farmers have also become recipients of WFP food. Now, communities of herders and nomads are agitating to be included, too.

"We also don't have enough to eat," says Ismail Burham Jibril Ishac, a leader of an Arab settlement who came to a Save the Children-U.S. nutrition and health center in the refugee camp near the town of El Geneina. "There's no affordable vegetables because of the bad harvest."

Mr. Ishac says his garden withered from a lack of rain. He says he normally had a year's worth of grain in his house, but most of that is gone and he must go to the market every day to buy food. Instead of giving leftovers to his goats and sheep, he now keeps the scraps for his family. Porridge from the day before dries in a pan set out in the morning sun. When a gust of wind topples the pan, the children scramble to gather the pieces of porridge and clean them. Mr. Ishac says it will be tonight's dinner.

"If there's no general distribution of food here," he says, "the prices will just keep increasing."

As will the hunger. With tens of thousands of refugees packed into the El Geneina area, the labor pool is bloated and wages are low. Several of the women in the weaving center also work at the local brick factory, where they form clay into rugged rectangles which harden in the sun. They say the pay is about 40 cents for every 1,000 bricks, which take about two days to make.

Only after 2,000 bricks will they have enough for a half-measure of sorghum at the market. That will make one meal.

DRY SPELL

Places can't stop drought from coming their way, but they can control its devastating effects.

BY CHRISTOPHER CONTE

Water officials in Denver didn't worry last April when the usual spring showers failed to materialize and people started watering their lawns two months ahead of schedule. After all, the city's reservoirs were more than three-quarters full, and one look at the snow-capped peaks of the Rocky Mountains seemed to promise a healthy spring run-off.

What they failed to check was the soil. Underneath the snow pack, it was seriously parched, and the mountain forests were desperately thirsty. When the spring melt began, the soil and trees drank up much of the water that would have filled streams in a normal year. May brought more trouble in the form of dry winds that vaporized much of the snow before it could melt, shrinking the snow pack to 13 percent of normal. Still, the water utility, blithely assuming reservoirs would refill as normal, merely recommended that people voluntarily cut their water use by 10 percent. Their light-hearted slogans—"Real Men Dry Shave" and "No Water, No Bikinis"—failed to stir much public concern.

It was only in July, when the flow of the south Platte River had slowed to less than 500 acre-feet of water compared with 30,000 normally and reservoirs had dropped to 60 percent of capacity, that Denver Water realized it had a crisis on its hands. In fact, it was confronting—and continues to confront—the worst drought in the central Rockies in 300 years.

The Mile High City is not alone in its distress. Normally, about 10 percent of the country suffers very serious drought at any time, but last summer the rate soared to 38 percent, and more than half of the country experienced abnormally dry conditions, if not outright drought. No region was immune. Some 18,000 private wells went dry in Maine. South Carolina, which went through its fifth consecutive year of dryness, was staggered by $520 million in timber losses due to slower tree growth and the loss of drought-weakened trees to pine beetles. Low river levels led to a surge in hydroelectric power prices in the usually rain-drenched Pacific Northwest. And in the Plains states, Kansas farmers last year reported $1 billion in crop losses due to lack of rain.

'Water is such a touchy subject here,' says Colorado's Brad Lundahl.

This year may not be much better. Agricultural analysts predict that the mild, dry winter in the heartland will lead to a plague of grasshoppers in every state west of the Mississippi River this summer. And the U.S. Army Corps of Engineers warned it might have to close this year's shipping season on the Mississippi and Missouri Rivers early because water levels may be too low for many barges.

While the recent drought has been, by many measures, as bad or worse than any in the past century, it isn't the first wake-up call to states that they can't afford to be lax about plans to deal with parched conditions. And there will be more warnings in the future. As the population continues to grow and shift both from rural areas to urban areas and from the humid East to the arid Southwest, water supplies are growing tighter. That suggests it may not take much of a deficiency in rainfall or snowfall to produce water shortages in the future. Add the possibility—unproven, but feared by many scientists—that global warming will lead to more frequent droughts, and you have a strong case that states—and localities—should get a lot more serious about dealing with drought.

Coming up dry

To their credit, states are becoming more aware of the risks. While just two states had formal drought plans in 1982, some 33 have such plans today, and seven more are working on the issue. But these plans generally deal with how governments will respond to a drought once its effects become apparent—an approach that falls far short of what hydrologists say is needed to avoid most of the damage. "If you wait until after a drought begins, all you can do is try to manage resources day to day," says Donald Wilhite, director of the National Drought Mitigation Center in Lincoln, Nebraska. "The time to prepare for a drought is before the drought begins."

Wilhite believes states have been shortsighted partly because the federal government is so quick to provide disaster-relief funds. That, in turn, diminishes the incentive to invest in preventive measures, such as monitoring conditions to detect when droughts might occur. "We have spent a lot of money putting out brushfires, and long-term monitoring has gotten the axe," says Barry Norris, the chief drought official for Oregon's Water Resources Department. For a capital investment of just $1.5 million, plus about $200,000 in annual operating expenses, Norris says he could install an effective network of equipment to monitor stream flows, snow pack and soil moisture levels. But in years when there is no drought, the expenditure for such a program can't compete with other state spending priorities. And when droughts occur, policy makers are too focused on short-term solutions to invest in policies that would pay off only in the long run.

Compounding the problem, state governments aren't set up to address drought issues comprehensively. In many cases, drought policy is the responsibility of agriculture departments, which in turn are beholden to farm constituencies that have grown adept at winning disaster-relief funds. In others, drought planning falls under the purview of emergency-preparedness agencies, which lack expertise in long-term water issues. And some state governments, especially in the West, feel thwarted by time-encrusted water-rights laws and powerful interest groups. "We could do more to be prepared for when the big droughts hit, but it's hard for me to get anything going," says Brad Lundahl, who as section manager for the Conservation and Drought Planning Section of the Colorado Water Conservation Board, is the state of Colorado's top drought official. "Water is such a touchy subject here, and the water community doesn't want to give us much authority."

The last straw

Denver's experience is a textbook example of complacency, the first phase of what hydrologists call a "hydro-illogical cycle" that characterizes government's traditional response to drought. Even though drought is as inevitable, if not predictable, as the change of seasons, governments tend to ignore the danger when water is abundant. As a result, they are slow to see it coming and ill-prepared for it when it arrives. Then, when they belatedly realize they are in a drought, their responses often are either ineffectual or even counterproductive.

"In retrospect, we should have been watching soil moisture and tree moisture, not just reservoir levels and snow pack," says Elizabeth Gardner, Denver Water's manager of conservation. "We also needed to make people recognize the seriousness of the drought earlier. And we needed a drought plan that was much more detailed."

In the hydro-illogical cycle, concern can quickly turn to panic in the absence of comprehensive drought planning. Unfortunately, panic rarely makes for good policy. In Colorado, for instance, drought has led officials to take a new look at sweeping proposals that had long been rejected as too costly or inimical to Coloradoans' sense of their own state. One proposal, dubbed the "Big Straw," would involve pumping water from the Colorado River near the Utah border back to the Continental Divide. The cost would be staggering: The state would have to build a 200-mile pipeline, at least two large reservoirs, a pumping system that would require the equivalent of 80 percent of the annual output of Hoover Dam to lift the water 4,000 feet over the Rockies, and new systems for cooling and purifying the water—all at a potential cost of $5 billion.

An even more draconian idea would involve clear-cutting large swaths of federal and state forests so that more snow would reach the ground, where, the theory goes, it eventually would melt and run into the streams that supply city reservoirs. "Logging for water," as that idea is known, could require cutting as much as 40 percent of the trees in watersheds, a process that not only would be expensive but also could damage natural habitats and do untold harm to the recreation and tourism industries.

Critics say such costs are not only destructive but also entirely avoidable. With prudent planning, they contend, a state such as Colorado can manage droughts for many years without taking such drastic and costly steps. "The cities of Colorado don't have to worry about running out of water for decades, or even centuries," says Douglas Kenney, research associate at the University of Colorado's Natural Law Resources Center. But, he adds, avoiding panic-induced measures will require policy makers to develop a whole new mind-set. "We have to stop thinking about drought as a phenomenon to be avoided at all costs, and think of it instead as a normal part of life," Kenney argues. "We have to get more used to the idea of risk management."

Fighting back

Risk management could be the byword of a new drought strategy that is taking shape in Georgia, where a broad-based working group has proposed a state drought plan that emphasizes permanent reforms rather than emergency responses. "We decided that a drought plan isn't about what you do in a drought, it's about what you do in advance to mitigate the effects of drought," says Robert Kerr, who, as director of the Pollution Prevention Assistance Division for the state Environmental Protection Division, led the planning effort.

133

Water Woes

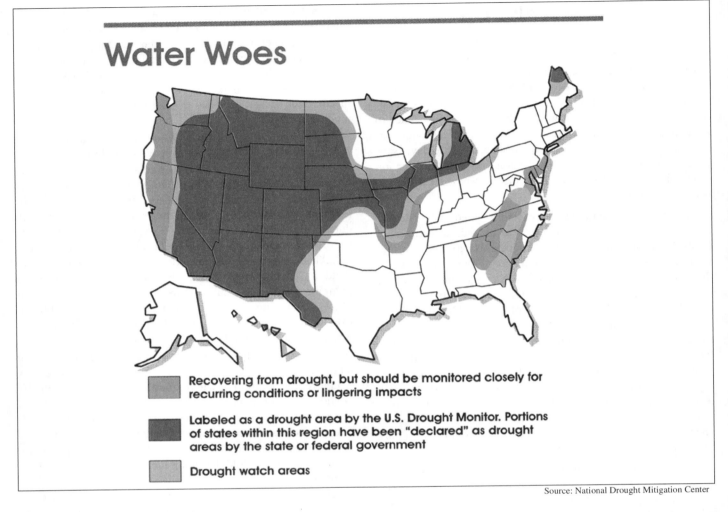

Recovering from drought, but should be monitored closely for recurring conditions or lingering impacts

Labeled as a drought area by the U.S. Drought Monitor. Portions of states within this region have been "declared" as drought areas by the state or federal government

Drought watch areas

Source: National Drought Mitigation Center

One key proposal, for instance, would regulate outdoor watering all the time—not just during droughts. The plan is controversial even though it would allow homeowners to water their yards three times a week, enough to keep any home landscape in good health. Indeed, it would make many yards healthier than they are today. That's because most homeowners tend to overwater their plants, a practice that prevents the plants from developing deep root systems that enable them to find water if the soil around them dries up. Rose Mary Seymour, an engineer and horticulturist at the University of Georgia, argues that modest restrictions on watering would both reduce the risk of drought by saving water and make droughts that do occur less damaging. Seymour works with landscapers and gardeners to encourage xeriscape, a system of water-efficient landscaping design and maintenance.

The focus on outdoor watering reflects a simple fact: For many municipal water systems, landscape irrigation represents the biggest, and probably least essential, category of water consumption. In Denver, for instance, outdoor watering accounts for 40 percent of all water use. But Georgia's growing commitment to water conservation involves other sectors as well. The state has been conducting water audits of major institutions, and the drought-planning group has studied ways to improve the efficiency of agricultural irrigation. In addition, the Metropolitan North Georgia Water Planning District, a state-created agency that charts water policy for municipal governments in the Atlanta region, is preparing a list of urban water conservation measures it believes could reduce water demand by as much as 15 percent—or 1 billion gallons a day—by 2030.

Among the most promising reforms: requiring that existing homes be retrofitted with newer, water-efficient shower heads and toilets; paying homeowners rebates for buying more efficient washing machines; mandating rain-sensors that turn off automatic watering systems when they aren't needed; requiring the use of water meters for multi-family as well as single-family homes; and doing water audits for homes, hotels and commercial establishments.

Of course, even aggressive conservation won't avert all droughts, Georgia officials concede. So their plan also calls for a sophisticated early warning system. Instead of relying on a single drought indicator, as Denver Water officials did last year, the Georgia plan will look at four indicators—stream flows, groundwater levels, reservoir levels and precipitation patterns. Anytime one or more of these indicators drops below a certain level in a particular climate zone, the state would order new restrictions on outdoor watering. And those limits would stay in effect until all of the "triggers" that indicate drought conditions have taken a turn for the better.

Authority figure

Georgia's proposals represent a significant increase in the role of state government, but officials believe they are necessary to help avoid problems such as those in Denver last summer. There, a plethora of independent municipal water systems all decided for themselves whether there was a drought, how serious it was and what should be done about it. Officials at Denver Water believe the mixed messages they sent the public are a major reason its own watering restrictions proved ineffective for six crucial weeks last summer. (Denver-area water managers subsequently began meeting to develop a more coordinated effort.) Georgia officials believe their approach is the right balance between state and local control: The state would set minimum requirements and municipalities would be free to adopt more stringent standards. "There needs to be a state coordinating mechanism, but it needs to be sensitive to the local context," says Anne Steinemann, a Georgia Institute of Technology drought expert who advised Georgia on its planning effort.

Taking Drought's Measure
The annual cost of drought:
$ 6 billion to $8 billion
The annual cost of flooding:
$ 5.9 billion
The annual cost of hurricanes:
$5.1 billion

Source: National Weather Service

The Georgia plan, however, also leaves some of the toughest decisions to local officials. For one thing, it gives them the responsibility to decide whether to reform water rates to promote conservation. Currently, many municipal water systems charge a flat rate regardless of how much water is consumed. Some even offer lower rates to heavy consumers. The lack of a clear price signal is one reason why some homeowners "turn their yards into rice paddies," says Roy Fowler, general manager of the Cobb County-Marietta Water Authority, a wholesale supplier for municipal water systems in the Atlanta area. Cobb-Marietta imposes a 25 percent surcharge on customers whose summer consumption exceeds 130 percent of their winter consumption. That, combined with such conservation measures as federally mandated water-efficient toilets, helped reduce per capita consumption among its customers from 146 gallons a day in 1990 to 130 gallons in 2000.

Fowler is one of a new breed of water managers. Traditionally, water managers defined their jobs mainly in terms of

supply—that is, they sought to provide as much water as consumers wanted, with no questions asked. But today, Fowler says, "managing demand" is just as important. The reason boils down to basic economics: Conservation saves money. "A conscientious conservation program," Fowler notes, "costs pennies on the dollar compared to digging a hole in the ground and calling it a reservoir."

Still, conservation creates a new kind of risk for water managers. As they wring the inefficiencies out of current urban water systems, they will have fewer options for finding easy savings in the future. Aware of this problem, they are starting to look for other ways to minimize the disruption they could face in the worst droughts. Robert Kerr, who is shepherding the Georgia plan through the regulatory process, believes the ultimate solution may be the water equivalent of "rolling blackouts," in which some industries might be willing to curtail operations on certain days in return for a price break on their water.

A shared solution

Long before that happens, cities could negotiate arrangements to tap into water generally used for agricultural irrigation during serious droughts. That represents a substantial potential reserve in most states. In Georgia, for instance, farm irrigation accounts for 30 percent of the water consumed during nondrought years. What's more, there are ample models for how such a water-insurance system could work: Federal farm policies have made farmers well accustomed to the idea of forgoing production in exchange for cash payments. Georgia already has applied the principle to water. In 2001 and 2002, both serious drought years, the state conducted auctions to buy back irrigation rights from farmers. Last year, the auctions took 42,000 acres out of irrigation at a cost of $5.5 million, an arrangement that added about 25 percent to the flow of the Flint River in the southern part of the state.

Water-sharing arrangements have even more potential in the arid West, where agriculture accounts for a larger percentage of total water consumption than in Eastern states such as Georgia. Indeed, some Colorado municipalities already have negotiated deals to lease farmers' water rights during emergencies, and environmental groups in the Rocky Mountain state are pushing legislation that would clear potential legal barriers to additional transfers. Environmentalists see the idea not only as an alternative to mega-projects like the "Big Straw" but also as a way to preserve open space. There's no question that cities have the economic clout to buy up all the rural water they need for the foreseeable future, the environmentalists note. But cities will buy less—and hence keep more land in agricultural production longer—if they can reach agreements that ensure the water will be available to them in emergencies.

Such stand-by arrangements may sound like a common-sense solution, but they aren't a sure thing politically. In Georgia, water auctions have proven to be controversial with city voters. "A lot of people say, 'I had to cut off watering my lawn and I didn't get paid. Why should farmers?'" notes Jim

Hook, a soil and water management specialist at the University of Georgia. "Farmers are such a small political constituency it's not clear there will be the political will to keep the water auctions up."

That points to perhaps the biggest challenge of all when it comes to planning for drought. Water issues are, to say the least, politically divisive. Many leaders are reluctant to take them on when there is no crisis to force the issue. Still, you don't have to be a weatherman to see that the issue can't be avoided forever. With continued population growth, conditions that are considered drought today could be the norm soon. Denver Water's long-range plans show, for instance, that the same degree of conservation the city ultimately achieved on an emergency basis last year—by fall, it had cut water consumption by 30 percent—will have to be a permanent way of life by 2050.

Will Denverites take the goal seriously once the snows return and refill their reservoirs? The eventual success of water restrictions last year offers reason to believe they will. But the hydro-illogical cycle suggests another outcome: a return to apathy. As John Steinbeck, chronicler of America's "Dust Bowl" drought years, noted, "It never failed that during the dry years people forgot about the rich years, and during the wet years they lost all memory of the dry years. It was always that way."

From *Governing*, March 2003, pp. 20-24. © 2003 by Governing. Reprinted by permission.

TURNING OCEANS INTO TAP WATER

DESALINATION PROMISES TO RESCUE SPRAWLING COMMUNITIES IN DIRE NEED OF FRESHWATER. IS THAT A GOOD IDEA?

TED LEVIN

America is running out of drinking water. In parts of the arid West, this is literally true. In coastal areas, such as Pinellas County, Florida, the problem more closely resembles Coleridge's famous verse, "Water, water, every where/ Nor any drop to drink." To slake its thirst, the local water authority, Tampa Bay Water, has built the largest desalination facility this side of Saudi Arabia. Situated on Apollo Beach, just across Tampa Bay from the Pinellas Peninsula, the plant is the only operational commercial desal facility in the United States. Eventually it will supply the region—a three-county area with more than two million people and growing—with 10 percent of its drinking water. (The rest will come from a now depleted aquifer, a new groundwater supply, and several aboveground rivers.)

The Apollo Beach plant may be a very good idea or a very bad one. It all comes down to this: Is desalination a legitimate response to a bona fide emergency, or is it simply an enabler for unchecked sprawl in fragile coastal areas that do not have the natural means to support their exploding populations?

Pinellas County, home of lovely St. Petersburg, is bounded on the west by the Gulf of Mexico and on the south and east by Tampa Bay. The soil is sandy and porous, perfectly suited for the engineering works of gopher tortoises. The beaches are classic Florida, bone-white sand lapped by blue water, beneath a wide arc of subtropical sky. In 1539, when Hernando de Soto marched up the Gulf coast, the Pinellas Peninsula was an open woodland of pines and palms and oaks. A dense coif of mangroves punctuated by salt marshes rimmed Tampa Bay, while the bay itself, covering nearly 400 square miles, was a mosaic of sea grass beds and oyster bars, mudflats and open water. In season, birds from across the continent convened in and

around Tampa Bay to gorge themselves on the flats and beaches and in the woodlands and shallows, where shoals of fish moved from the Gulf to spawn or feed in the fecund estuarial waters. Sea turtles nested on the beaches. Manatees grazed the sea grass beds. Back then, before the dredging of shipping lanes, a man could have threaded his way across the shallow bay without wetting his hair.

Tampa Bay remained a symphonic wilderness well into the nineteenth century, but its despoliation was swift. In the late 1880s, the hub of Pinellas County was an unnamed community, population 30. In 1892, the community incorporated into St. Petersburg, population 400. Early in the last century, to meet future water needs, Pinellas County and the city of St. Petersburg bought land in the hinterlands of Pasco and Hillsborough counties, north of Tampa Bay. Eleven well fields set in remote wetlands supplied the city with the potable ground-water that the peninsula itself could not provide.

By 1920, the population of Pinellas County had reached 28,000. Five years later, after a six-mile bridge was built to connect Pinellas County and Tampa, the population had grown to 50,000. By 1950, it was 159,000. By 1970, it had soared to 522,000. Today, as Pinellas County's population reaches nearly a million, Pasco and Hillsborough counties have undergone population explosions of their own, further stressing the well fields. Surrounding wetlands have become fire hazards and nearby lakes have receded from their shores. The faucets of some Pasco County residents literally have run dry.

A century of dredging, filling, building, and digging has destroyed 80 percent of the sea grass beds and more than 40 percent of the mangroves and salt marshes. Storm water runoff from cities and farms and the dumping of untreated sewage continue to strangle Tampa Bay. Nitrogenous compounds from coal-fired power plants and automobile exhaust fall out of the air, lacing the rain with toxins and turning the bay's gin-clear water into an opaque algal soup that has smothered the sea grass beds.

Only 3 percent of the earth's water is fresh, and more than two-thirds of that is bound up in glaciers and ice caps, rock-hard and beyond reach. This leaves less than 1 percent of the planet's water available for drinking and washing and mixing with bourbon, and that meager amount is not evenly distributed.

On the face of it, the Tampa Bay region would seem to have an abundance of aqueous resources. Buried among the layers of sedimentary rock beneath Florida and its continental shelf lies an ancient bubble of freshwater, the Floridan Aquifer, one of the largest in the world. Like the state, the aquifer is bounded on three sides by salt water. The layered rocks hold roughly two quadrillion (that's 2,000,000,000,000,000) gallons of water. To this hefty amount add 50,000 miles of rivers and streams, nearly 8,000 lakes and ponds, and 600 springs, some so large they become navigable rivers when they reach the surface. All this water sits on, or under, or slices through, more than three million acres of wetlands. When compared to other Sun Belt states, Florida appears submerged in good fortune. The question arises, then: Why are the 11 well fields that serve the greater Tampa Bay area running out of water?

One reason is that groundwater does not behave like surface water. Wells take longer than lakes to recharge, and the lower pressure created by depleted wells pulls surface water downward. The more water drawn out of a well field, the deeper and wider the zone of lower pressure, and the more surface water fills the void. As surface water drains away, wetlands dry out, and even though particular localities sit atop a subterranean sea of freshwater, they may suffer a dramatic loss.

Prior to the passage of the state's 1972 Water Resources Act, which established five regional water management districts within the Florida Department of Environmental Protection, anyone could drill anywhere. After 1972, the water management districts began to issue consumptive use permits. Twenty years later, when Pinellas County's groundwater permits expired and Pasco County balked at having them renewed, the crisis moved from the faucets to the courts, eating up millions of dollars in legal fees.

In 1997, after a lengthy and contentious review process, the South-west Florida Water Management District agreed to cofund a search for new supplies of freshwater for the Tampa Bay area. In an effort to alleviate Pasco County's water shortage, the water management district agreed to scale back pumping of the well fields. The goal was to reduce the level of pumping by more than half—from 192 million gallons a day (mgd) in 1996 to an eventual low of 90 mgd by 2008. This reduction, hydrologists hoped, would be enough to restore the health of the aquifer. By 1998, continued water shortages forced the governments of Hillsborough County, Pasco County, Pinellas County, St. Petersburg, New Port Richey, and Tampa to try something new. They decided to commission the construction of what would be the largest desalination plant in the country.

Until very recently, the notion of drinking seawater was lunatic fringe, involving a technology suitable for nuclear submarines and the Middle East, where an oil-rich, water-poor landscape makes financial and practical obstacles irrelevant. In 1960, there were just five desalination plants worldwide. Until the late 1990s, only two American cities had invested in full-fledged desal plants—Key West, Florida, in the 1980s, and Santa Barbara, California, a decade later. Both cities shelved their plans soon after the facilities were built, having found less expensive sources of water elsewhere. It is still cheaper for Key West to pump freshwater 130 miles from beneath the apron of the Everglades than to desalinate seawater.

However, as desalination technology improves, lowering the cost of producing freshwater, more planners are looking to the ocean as the droughtproof guarantor of continued growth. Throughout the Sun Belt, metastasizing communities have outstripped existing water supplies and begun to look seaward. Last year, municipal water agencies from California, Arizona, New Mexico, Texas, and Florida pooled resources and formed the U.S. Desalination Coalition, a Washington, D.C.-based advocacy group that lobbies the federal government to invest in new desalination projects.

Today there are more than 12,500 desal plants in 120 countries, mostly in the Middle East and Caribbean. Saudi Arabia meets 70 percent of its water needs by distilling salt water; the British Virgin Islands Tortola and Virgin Gorda rely on desalination for 100 percent and 90 percent of their respective water needs. The American Water Works Association, the largest organization of water professionals in the world—its 4,500 utility members serve 80 percent of America's population—forecasts that the world market for desalinated water will grow by more than $70 billion in the next 20 years.

California will soon be in the vanguard in the United States. It has already planned or proposed about a dozen desal plants along its coast, including a $270 million plant in northern San Diego County slated for completion in 2007. Early last year, the federal government reduced the amount of Colorado River water allocated to Southern California, forcing the state to accelerate its search for alternative sources after years of helping itself to the dun-colored Colorado at the expense of other western states (and Mexico).

To learn about the potential impact of desalination, I visit Mark Luther at the University of South Florida's Marine Science Center, in St. Petersburg. After a slow drive across the Pinellas Peninsula, traffic congealing at every intersection, I pull into the science center parking lot. It's an early December afternoon, hot and dry, the sky blue from seam to seam. High above the lot, an osprey throws a tantrum, lobbying for issues beyond my comprehension. From the second floor of the building I can see the desal plant across the bay on Apollo Beach, white like the salt it removes. Luther is the oceanographer who studied the bay's circulation patterns as part of the environmental assessment team that helped Tampa Bay Water determine where to site the facility. We settle at a black laboratory table in his bright, cluttered office. Luther, 50, wears a powder-blue yacht club T-shirt and sockless moccasins. His eyes match his shirt. His sand-colored, shoulder-length hair hangs in a ponytail. Luther tells me that, on average, 60 cubic meters of freshwater a second flow into the head of Tampa Bay, courtesy of four main rivers—the Hillsborough, the Alafia, the Manatee, the Little Manatee—and a number of smaller tributaries. The freshwater, lighter than salt water, is stirred by the tides before draining into the Gulf of Mexico.

"No matter where you take freshwater, it's going to have some impact on the environment," Luther says. "The goal is to distribute the sources to reduce that impact." Besides operating the desal plant, Tampa Bay Water pumps two new groundwater sites and diverts water from three of the rivers that feed Tampa Bay. "Taking river water has a much larger impact on the bay than the desalination facility," he says. "Of all the ways to get potable freshwater, building a desal plant is no worse and probably better than overpumping well fields or diverting too much river water." It's hardly a ringing endorsement, but it also suggests that an intelligently planned desal plant is not something a sensible environmentalist should lose too much sleep over.

You can't locate a desalination plant just anywhere, however. You need an energy source to operate the plant and a circulation pattern that removes the discharged brine. Brackish water, being less salty than seawater, costs less to desalinate. Hence, the plant was built inside the bay, on Apollo Beach, where salinity, though varying seasonally, averages 20 parts per thousand (ppt), 15 ppt lower than in the Gulf of Mexico. The Big Bend coal-fired power plant sits next door, providing a ready source of water and energy. Of the 1.4 billion gallons the power plant uses each day to cool its condensers, Tampa Bay Water recycles 44 million gallons for desalination. Because the plant already passes intake water through a pair of screens to filter out fish and other sea organisms, from fish eggs to plankton, the desal facility does not cause any additional loss of aquatic life. From the 44 million gallons of salt water it receives daily, the plant produces 25 million gallons of freshwater. The highly concentrated salt water that remains is mixed with the power plant's effluent before being returned to Tampa Bay.

This discharge water adds only marginally to the salinity of the bay, says Luther. A little more than a quarter of a mile from the discharge site he could not detect any increase in salinity. "We're at least an order of magnitude less than natural variability," he reports. The circulating currents and tides, aided by a 43-foot-deep shipping lane dredged decades ago, wash the brine away from Apollo Beach.

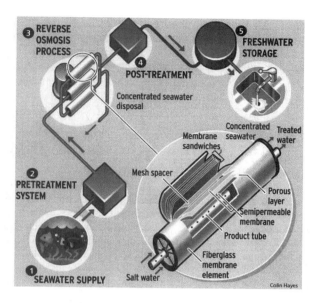

Colin Hayes

Put simply, desalination purifies water by removing dissolved mineral salts and other solids. In the Middle East, most desalted water is produced by means of distillation, which imitates the natural water cycle: Salt water is heated to produce water vapor, which then condenses to form freshwater. American desal plants favor a different technology—reverse osmosis—which forces the water through a series of membranes, leaving the salts behind. Tampa Bay Water engineer Ken Herd, top, shows a cutaway model of one section of wound membrane. Hollow at its core, each section conveys the desalination water to the final "post-treatment" phase.

Not everyone believes the desal plant is benign. According to an advocacy group called Save Our Bays, Air and Canals (SOBAC), which has its headquarters in Apollo Beach, Tampa Bay takes two years to flush. The briny discharge, SOBAC claims, is equivalent to dumping a truckload of salt in the bay every 16 minutes. The group says that part of the littoral zone off Apollo Beach is already hypersaline. Luther does not believe the desal plant will add to the problem. This part of Tampa Bay flushes about every two weeks during the summer, he tells me, less frequently during the winter. "The waters off Apollo Beach are constantly refreshed. That's why the site was chosen.

"It's ironic that SOBAC brings up hypersalinity," Luther adds. "Probably the biggest environmental disaster to hit Tampa Bay in the last 50 years was the construction of the Apollo Beach community. They dredged pristine mangroves and sea grass beds to build stagnant finger canals and spits of land that are now heavily developed. All those waterfront homes have nice green sodded lawns that require fertilizers and pesticides, which drain right into Tampa Bay."

As a naturalist, I know that filtering salt from seawater is not a novel idea. For hundreds of millions of years marine plants and animals have evolved unique methods of desalination. Salt glands discharge excess salt through the nostrils of marine iguanas, the eyes of sea turtles, and the tongues of crocodiles. The underside of the leaves of black mangrove trees exude pure salt crystals that glisten in the tropical sun; the spidery roots of red mangroves block salt from entering the tree. The gills of saltwater bony fish such as tuna or striped bass, the rectal glands of sharks and rays, and the super-kidneys of whales and seals perform a similar function.

I want to understand how desal works for humans, so I drop in on Ken Herd, 43, engineering and projects manager at Tampa Bay Water's Clearwater office complex. Tampa Bay uses a reverse osmosis (R.O.) membrane system, explains Herd, in which salt water is pushed at extreme pressure, up to a thousand pounds per square inch, through tiny pores, each 0.0001 micron in diameter—approximately 1/1,000,000 the width of a human hair.

Osmosis, as you may recall from 10th-grade biology, is the tendency of a fluid to pass through a semipermeable membrane, such as the wall of a living cell, into a solution of higher concentration, to equalize concentrations on both sides of the membrane. Reverse osmosis is precisely … the reverse. The pores of the roughly 10,000 tightly rolled membranes are so small that ultratiny molecules of water pass through, but larger molecules of dissolved minerals like salt do not. Pressure forces out the salt, and the constant flow of water helps wash the outer membranes clean of concentrations of brine. R.O. membranes still clog, however, and have to be cleaned, every three weeks to six months or longer. The membranes last five to seven years, sometimes ten, and they are expensive to replace.

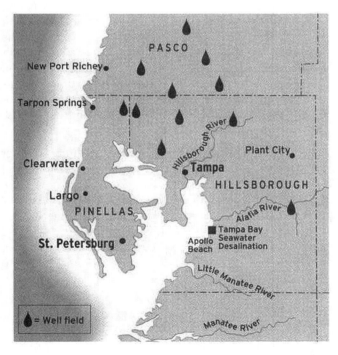

Steve Stankiewicz

Herd shows me a model of a three-foot section of wound membrane. It looks like an oversize roll of paper towels, with the top cut away so that I can see inside. Salt water forced against the outside of the roll filters through the spiral until pure freshwater flows into the center port—the equivalent of the cardboard tube inside the roll of paper towels—and then out into a network of collecting pipes. The total surface area of the plant's 38-inch-wide membranes would cover nearly 65 football fields.

"However," says Herd, "R.O. is the simplest part." First, the bay water must be treated before it's forced across the R.O. membranes. Pretreatment filters out suspended solids—such as scraps of seaweed, fish fry, aquatic larvae, sundry items of flotsam and jetsam. If this weren't done, the membranes would foul. "Pretreatment," says Herd, "is the challenging phase of desalination." Tampa Bay Water uses dual-stage sand filtration, in which incoming salt water flows up through two filtration cells, coarse-and fine-grained. Particulate matter larger than five microns in diameter that manages to pass through the sand filters gets caught in the cartridge filter—a collection of long, thin filters, like those used in swimming pools, which act as the R.O. membranes' safety net.

In every performance test, both the pretreatment filters and the R.O. membranes clogged more frequently than expected, requiring additional cleaning. Increasing the strength of the cleaning solution for the membranes caused another unforeseen problem: Two million gallons of caustic, soapy cleaning fluid had to be transported to Tampa's wastewater treatment plant.

Asian green mussels turned out to be the culprit. The alien shell-fish first appeared in Tampa Bay about eight years ago, having been transported in the ballast of tankers, and has thrived. Mussels love flowing, food-rich water, so the power plant's daily 1.4 billion gallons of effluent is bivalve utopia. Larvae pass through the power plant's intake screens, survive in the heated water, then clog the pretreatment filters, fouling the R.O. membranes with microscopic hairs.

The post-treatment phase also has its complications. Along with salt, alkaloids are stripped out of the water, leaving the desalinated water acidic and corrosive to pipes. So calcium carbonate (lime) is added during post-treatment, raising the pH level before the water is piped 14 miles to storage. All this trouble and delay has resulted in lost time and money. The desal plant has declared bankruptcy three times, most recently in October 2003. The plant is online only once a month, and Tampa Bay Water says it will not go into full production until 2006.

Contemplating the sprawl that surrounds the Apollo Beach plant, I find myself paraphrasing the line from the Shoeless Joe Jackson character in the movie *Field of Dreams*: "If you build it, they will come." Herd bristles a little at the phrase. "The government agency that allows growth supports its decision with electricity, drinking water, and waste removal. The water management district doesn't have the ability to limit growth; that's the job of the planning board. Tampa Bay Water just supports the growth that's already there."

Tampa Bay Water admittedly has taken significant steps to diversify its sources of potable water, and to do so in an environmentally responsible way. As of April 2004, the water authority was pumping only 74 million gallons a day from the ailing well fields, in hopes of restoring that corner of the Floridan Aquifer. As a result of these reductions, the surrounding wetlands have begun to recover—lake levels are rising and marshland vegetation is looking fuller, more lush, a de Soto shade of green. "We didn't trade one environmental impact for another in Pasco County by shifting the burden to Tampa Bay," Herd says with justifiable pride.

Herd's optimism is refreshing. And he's right: It is not ultimately the water authority that determines the carrying capacity of a suburban landscape. Many of the 20 commercial seawater desalination projects under consideration for the Sun Belt are driven by planners who both forecast and encourage growth, often in ecologically sensitive coastal areas. Faced with lobbying by the U.S. Desalination Coalition, environmentalists will need to scrutinize each new project. For if new desal facilities mean that the wild hills become crowded with condos and the shorelines fill with sprawl, we may find ourselves echoing another line that's associated with the hero-turned-villain of the 1919 Black Sox scandal. We'll have built it, they'll have come, and like the distraught young fan, we'll be exclaiming, "Say it ain't so, Joe."

Putting the world to rights

COPENHAGEN

What would be the best ways to spend additional resources on helping the developing countries? Some answers

IN RECENT weeks *The Economist* has been following and supporting the Copenhagen Consensus project—an unusual, ambitious and, some have argued, misguided attempt to set priorities among a range of ideas for improving the lives of people living in developing countries. Starting on April 17th, we began publishing, both in print and on our website, reviews of essays commissioned by the organisers from leading economic researchers. Each of the papers addressed one of ten global challenges, and proposed possible responses. During May 24th-28th, a panel of distinguished economists assembled in Copenhagen. Their task was to review these papers alongside critical commentaries commissioned from other researchers, to question the various authors, and to decide what to make of it all.

The organising idea was that resources are scarce and difficult choices among good ideas therefore have to be made. How should a limited amount of new money for development initiatives, say an extra $50 billion, be spent? Would it be possible to reach agreement on what should be done first?

The drive behind this venture was supplied by Bjorn Lomborg, author of that modern classic of green de-mythology, "The Skeptical Environmentalist". Mr Lomborg is a figure of controversy around the world and especially in his native Denmark, where he is currently head of the Environmental Assessment Institute. The meeting was hosted by the institute, with the support of the Danish government.

Mr Lomborg's role, and the somewhat hubristic character of the undertaking (who do these economists think they are, and when have economists ever agreed about anything?), helped to draw attention to the event. So did the calibre of the various challenge-paper authors and discussants, and of the "expert panel" that met to judge their submissions. This panel of eight included three Nobel prize-winners—Robert Fogel of the University of Chicago, Douglass North of Washington University in St Louis, and Vernon Smith of George Mason University. And the other five, who may collect a few more Nobels in due course, are also eminent in their respective fields—Jagdish Bhagwati of Columbia University, Bruno Frey of the University of Zurich, Justin Yifu Lin of Beijing University, Thomas Schelling of the University of Maryland, and Nancy Stokey of the University of Chicago.

At an earlier stage, the panel had narrowed a much larger number of development challenges, drawn from assessments of the United Nations and its agencies, down to just ten:

- Civil conflicts
- Climate change
- Communicable diseases
- Education
- Financial stability
- Governance
- Hunger and malnutrition
- Migration
- Trade reform
- Water and sanitation

That list itself was somewhat controversial, mainly because of what it left out. ("Why nothing on the role of women?" the panel was repeatedly asked by the press during the course of the week.) But the list of ten challenges will strike many as less surprising than the prioritised list of policies that the panel then drew up.

The top of this list (see table) was

The results

Project rating		Challenge	Opportunity
Very good	1	Disease	Control of HIV/AIDS
	2	Malnutrition	Providing micro nutrients
	3	Subsidies and trade	Trade liberalisation
	4	Disease	Control of malaria
Good	5	Malnutrition	Development of new agricultural technologies
	6	Sanitation and water	Small-scale water technology for livelihoods
	7	Sanitation and water	Community-managed water supply and sanitation
	8	Sanitation and water	Research on water productivity in food production
	9	Government	Lowering the cost of starting a new business
Fair	10	Migration	Lowering barriers to migration for skilled workers
	11	Malnutrition	Improving infant and child nutrition
	12	Malnutrition	Reducing the prevalence of low birth weight
	13	Diseases	Scaled-up basic health services
Bad	14	Migration	Guest-worker programmes for the unskilled
	15	Climate	"Optimal" carbon tax
	16	Climate	The Kyoto protocol
	17	Climate	Value-at-risk carbon tax

Source: Copenhagen Consensus
Note: Some of the proposals were not ranked

not the problem. Ranked first was a package of measures aimed at controlling HIV/AIDS. Next came a set of interventions aimed at fighting malnutrition. The third-ranked policy did raise a few eyebrows, among economists and NGO sceptics alike: "multilateral and unilateral action to reduce trade barriers and eliminate agricultural subsidies." ("Why so low?" ask economists. "How come so high?" reply the NGOS.) In fourth place, also unlikely to arouse much protest, were new measures to control malaria.

The panel rated all four of those proposals "very good", as measured by the ratio of social benefit to cost. In fact, by the ordinary standards of project appraisal, they are not just very good but extraordinarily good, with benefits exceeding costs by a factor of ten or more, and sometimes

much more. That proposals this good should fail to be adopted for lack of finance is a scandal, especially when you reflect on some of the projects that governments are currently financing.

The bottom of the list, however, aroused more in the way of hostile comment. Rated "bad", meaning that costs were thought to exceed benefits, were all three of the schemes put before the panel for mitigating climate change, including the Kyoto protocol on greenhouse-gas emissions. (The panel rated only one other policy bad: guest-worker programmes to promote immigration, which were frowned upon because they make it harder for migrants to assimilate.) This gave rise to suspicion in some quarters that the whole exercise had been rigged. Mr Lom-

borg is well-known, and widely reviled, for his opposition to Kyoto.

These suspicions are in fact unfounded, as your correspondent (who sat in on the otherwise private discussions) can confirm. A less biddable group would be difficult to imagine. The challenge-paper on climate change was written by William Cline of the Centre for Global Development; Mr Cline is pro-Kyoto, and in fact favours even stronger measures to abate carbon emissions than the protocol requires. But the panel insisted on making their own minds up on the issue. Right or wrong, there was no dissent among any of the eight.

Interestingly, an invited gathering of young people from around the world attending a "youth forum" run in parallel with the main event, and hearing the same submissions from challenge-paper authors and their discussants, ranked climate change only ninth out of the ten global challenges. So much for the view that age blinded the expert panel to the long-term dangers of global warming ("they won't be around to suffer the consequences"). Perhaps this should give pause to governments dedicated, or claiming to be dedicated, to Kyoto's implementation.

Who do they think they are?

How valid was this exercise even in economic terms—to say nothing of whether economists should give advice on choices in which ethics and politics play a great part? So far as the economics goes, the main issue is whether the proposals can properly be compared with one another.

A partial answer is that they will be so compared—if not explicitly, then implicitly—whether economists like it or not. Governments do in fact set priorities among the well-intentioned projects they can choose to finance, merely by deciding to do some things and not others. A ranking could be inferred from the choices they actually make. So why not face the question of priorities head-on? If it has to be done, it is best to do it well—and that, despite the undeniable difficulties, means giving the issue of priorities some thought.

Cost-benefit analysis must be the organising method for any such analysis—even if one thinks of this approach as inconclusive on its own, or as little more than a way of ordering one's thoughts more logically. Trying to calculate costs and benefits has the virtue of forcing assumptions into the open. If it turns out that different projects are being compared on the basis of different assumptions about, for instance, the value of a life saved, or the rate at which future costs and benefits should be discounted, it will become obvious, and allowance can be made.

And this, in fact, is exactly what the expert panel found. The challenge-paper on climate change, for instance, proposed the use of a far lower discount rate than the other studies (though it offered reasons for doing so). And estimates of the value of life (or of a "disability-adjusted life-year", the standard yardstick) likewise tend to vary from study to study. The panel allowed as best it could for differences such as these. It also had to take two other broad points into consideration.

One is that there is a great deal of interdependence among the proposals. For instance, one of the challenge areas was civil conflicts. Several proposals for reducing their frequency, duration and severity were discussed. If such measures were to succeed, however, the benefits would extend far beyond reducing direct losses in lives and livelihoods. The effectiveness of many other forms of aid depends on civil order. In countries riven by conflict, little can be done to speed economic progress. Given peace and security, well-directed aid has a chance of working. In principle, the calculation of costs and benefits should take such linkages into account. In the case of civil conflicts, the implication is that the benefits that would flow from reducing their incidence are much larger than they might at first appear.

A second difficulty, frequently emphasised by members of the panel during their discussions, is that the likely effect of most if not all of the interventions under review depends on the local institutional context. For instance, spending more money on education may yield good results in countries where schools are already held accountable, in one way or another, for the quality of the service they provide. Where they are not held accountable, which is the position in most of the least developed economies, additional spending may simply be wasted.

Since the institutional setting varies greatly from country to country, generalising about the prospects for any particular policy is dangerous. Throughout, the panel leaned towards recommendations which made the smallest demands on government agencies and other institutions—that is, towards proposals where there seemed less to go wrong. If one could assume "good government", many of today's intractable development problems would be soluble. Needless to say, one can make no such assumption.

Even if these questions of interconnectedness and institutions could be waved away, and basic parameters such as the value of life and the discount rate were clear-cut, the cost-benefit approach would still be difficult to implement. After all, it requires forecasts of costs and benefits, often extending far into the future—in the case of global warming, literally centuries into the future. Such forecasts will always be prone to error. And none of the panel was willing to let a mechanical assignment of costs and benefits drive the process to the exclusion of everything else. Their concluding statement emphasised that weight had been given as well to "the demands of ethical or humanitarian urgency".

All those in favour

You might think this would have condemned the Copenhagen Consensus project to irrelevance—or at any rate to inconclusive squabbling among members of the panel. Surprisingly, that did not happen. Admittedly, sceptics could argue that the meeting had been designed to yield an apparent consensus even where there was none: the group's ranking of proposals was to be derived by taking the median of all the members' individual rankings, assuring a result, at least in an arithmetical sense. Yet each member was still going to have to endorse the collective ordering—which they did, and which was not to be taken for granted.

Also, since the individual rankings are to be published in due course, along with members' commentaries explaining their choices, it was desirable that the different orderings should look quite similar. Otherwise, the result would command little respect. In fact, the individual orderings were pretty much alike. Despite clear differences of intellectual approach and of ideological tendency, the group did indeed arrive at a broad consensus—both in setting priorities and in agreeing that some projects could not be ranked.

Altogether the challenge-paper authors offered 38 proposals for action. The panel chose to rank only 17 of these, deeming that for the other 21 there was too little information to make a clear judgment about the relative merits. (A proposal was included in the group ranking only if five of the members had included it in their individual ranking. Again, however, there was surprisingly broad agreement about which proposals to rank and which not. With only one or two exceptions, policies tended to be ranked either by all of the members or by none.)

With something close to unanimity, the panel put measures to restrict the spread of HIV/AIDS at the top of the ranking. The challenge paper on communicable diseases, by Anne Mills and Sam Shillcutt of the London School of Hygiene and Tropical Medicine, having reviewed the literature, reckoned that a package of preventive measures costing some $27 billion (in purchasing-power-adjusted dollars) over eight years would prevent nearly 30m new infections (reducing expected infections from 45m over the period to 17m). One study has calculated that part of this package, condom distribution combined with treatment for sex workers who are suffering from sexually-transmitted diseases, would entail a cost of just $4 for each disability-adjusted life-year saved. The implied ratio of benefits to costs is nearly 500—and this is assuming a value of life, based on GDP per head, that is significantly lower than the figure of $100,000 which the panel said it preferred to apply.

Not only are millions of lives directly at stake. In sub-Saharan Africa, the toll of AIDS is so terrible that whole societies are in danger of breaking down. Despite the fact that the issue has received enormous attention of late, efforts to remedy the problem are still curtailed by lack of funds. The daunting scale and urgency of the issue, no less than the estimated costs and benefits of prompt action, persuaded the panel to make this its highest priority.

The same paper made a similarly compelling case for more to be spent on the control and treatment of malaria. Several different interventions were recommended, notably the wider provision of bednets treated with insecticide. Distribution of nets does require a basic health-service network, but experience in Tanzania, for instance, suggests that coverage of the population can be substantially increased given extra resources. Benefit-cost ratios are high. The panel earmarked more than $10 billion of its hypothetical $50 billion for this and a range of other anti-malaria initiatives, putting the package as a whole at number four in the ranking.

In second place, just behind control of HIV/AIDS, came a proposal to attack malnutrition—iron-deficiency anaemia, in particular—through a targeted programme of food supplements. Again the evidence suggests that the idea is feasible, and that it offers exceptionally high ratios of benefits to costs. This would account for the remainder of the putative $50 billion.

What about trade reform? The panel was keen on it, as you might expect, but the issue caused more disagreement than any of the other top four recommendations. The net global benefits of free trade, including the elimination of agricultural subsidies, would be enormous, according to the challenge-paper by Kym Anderson of the World Bank, maybe running into the trillions of dollars. According to one view, the

budgetary outlays needed to secure these benefits are actually zero, implying not just a huge flow of net benefits but also a benefit-cost ratio of infinity. Beat that.

But two cautionary notes were entered. First, eliminating farm subsidies, often demanded of rich-country governments as a pro-poor policy, would, at least in the first instance, hurt some developing countries: the ones, often the very poorest, that are and expect to remain net importers of food. A way must be found to help them. Second, trade liberalisation hurts some workers even in rich countries. More generous trade-adjustment assistance may therefore make sense, especially if it is aimed at workers displaced from industries struggling to survive in any case.

Proposals for spending more on water and sanitation were approved, and ranked high, in places six to eight inclusive, with little to choose among them. No education projects were ranked. Nor was Barry Eichengreen's intriguing proposal (see our Economics focus of April 17th) for the fostering of new bond markets. Nor were any proposals for better governance, except for the proposal to lower state-imposed costs on new businesses, which got the nod because the costs are low, the institutional requirements modest, and the possible benefits very great. In all these cases, the panel reckoned there was too little research to go on.

At the foot of the list stand the three proposals on global warming. All require sharp reductions in carbon emissions starting soon, reflecting the view of the challenge-paper author, William Cline, that bold action on the problem is warranted, and quickly. The panel, all in agreement, simply refused to buy it. The issue is real, they said, but not so urgent that such massive abatement costs need to be incurred right now. One of the commentaries on Mr Cline's paper, by Robert Mendelsohn of Yale University, proposed starting with a much lower carbon tax than implied by Mr Cline's three variants—at say $2 a tonne (compared with $150 in Mr Cline's "optimal" carbon-tax plan), rising in later years as more information on both the hazards and the technological opportunities became available. The panel thought that was more like it.

So the Copenhagen Consensus ended, surprisingly enough, in consensus. Mr Lomborg is again to be congratulated for his intellectual entrepreneurship. If rich-country governments want excellent value in return for an increase in their taxpayers' dollars spent confronting global challenges, they could do a lot worse than look closely at the highest-ranked ideas from this exercise.

MEXICO: Was NAFTA Worth It?

A tale of what free trade can and cannot do

By Geri Smith and Cristina Lindblad

Piedad Urquiza probably doesn't know much about NAFTA, but she knows what it's like to have a steady job. Urquiza works at a Delphi Corp. auto-parts plant in Ciudad Juárez, just across the border from El Paso. The assembly line is a cross section of working-class Mexico, from twentysomethings raised in this border boomtown to veteran hands harking from the deep interior. In the years since NAFTA lowered trade and investment barriers, Delphi has significantly expanded its presence in the country. Today it employs 70,000 Mexicans, who every day receive up to 70 million U.S.-made components to assemble into parts. The wages are not princely by U.S. standards—an assembly line worker with two years' experience earns about $1.90 an hour. But that's triple Mexico's minimum wage, and Delphi jobs are among the most coveted in Juárez. "I like the environment, I like my colleagues," says Urquiza, a 56-year-old widow who assembles the switches that control turn signals. The daughter of a poor rancher, she dropped out of school after the seventh grade and has relied on her Delphi job to raise six children to adulthood—and, she hopes, to a better life.

The pact is one of the **BIGGEST**, most **RADICAL** trade experiments in history

Urquiza and millions of other Mexicans live out daily one of the most radical free-trade experiments in history. The North American Free Trade Agreement ranks on a par with Europe's creation of the euro and China's casting off Marxism for capitalism. It encompasses 421 million people and melds two first-world economies—the U.S. and Canada—with a struggling third-world country, Mexico. The bloc was seen as a bold attempt to demonstrate once and for all the free trade's vast power to turn a developing nation into a modern economy. If anything was a litmus test for globalization, NAFTA was it.

MEXICO Then & Now

Despite doubts among Mexicans, the benefits under NAFTA are numerous

1993	2003
Government	
Single-party dominated	Multiparty democracy
Gross domestic product	
$403 Billion	$594 Billion
Exports as % of GDP	
15%	30%
Oil as % of GDP	
18%	9%
Remittances by migrants in U.S.	
$2.4 Billion	$14 Billion

Data: Mexican Central bank, Economist Intelligence Unit

PROMISES, PROMISES

ON JAN. 1, NAFTA will celebrate its 10th anniversary. The assessment? The grand experiment worked in spades on many levels. American manufacturers, desperate for relief from Asian competition, flocked to Mexico to take advantage of wages that were a 10th of those in the U.S. Foreign investment flooded in, rising to an annual average of $12 billion a year over the past decade, three times what India takes in. Exports grew threefold, from $52 billion to $161 billion today. Mexico's per capita income rose 24%, to just over $4,000—which is roughly 10 times China's."NAFTA gave us a big push," Mexican President Vicente Fox told *BusinessWeek*. Fox notes proudly that Mexico's $594 billion economy is now the ninth-largest in the world, up from No. 15 a dozen years ago. "It gave us jobs. It gave us knowledge, experience, technological transfer."

Just as important, the pact spurred profound political change. Mexicans who backed open markets also wanted an open political system. Would the Institutional Revolutionary Party (PRI) have fallen from seven decades in power in 2000 if Mexico hadn't signed a treaty requiring government transparency, equal treatment for domestic and foreign investors, and international mediation of labor, environmental, and other disputes? It's hard to believe democracy would have come to Mexico as quickly without NAFTA.

Impressive milestones—and seemingly ample proof that free trade delivers the goods. But rightly or wrongly, a large proportion of Mexicans today believe the sacrifices exceeded the benefits. The Mexican mood is infecting other Latin countries, which after 15 years of gradually opening their own economies to trade and investment are showing pronounced fatigue with the "Washington consensus," the free-market formula preached by the U.S. and the International Monetary Fund. In an August poll of 17 Latin countries carried out by Chile-based Latinobarómetro, just 16% of respondents said they were satisfied with the way market economics were working in their countries. Thus NAFTA's perceived shortfalls are giving fresh ammunition to free trade's opponents. "Now you have a whole network of people organizing against the Free Trade Area of the Americas and globalization because of what has happened in Mexico under NAFTA," says Thea Lee, the AFL-CIO's chief expert on international trade pacts. That's an ironic twist: It was NAFTA, after all, that kicked the global free-trade movement into high gear, spurring forward the Uruguay round of global trade talks in the mid-1990s and setting the stage for China's entry into the World Trade Organization in 2001.

Why have so many Mexicans soured on NAFTA? One problem is that the deal was oversold by its sponsors as a near-magic way to turn Mexico into the next Korea or Taiwan. Ten years later, many think the pact has stopped paying dividends—and that Mexico has been unfairly neglected by a Washington consumed by the war on terror. Speaking before an audience of Mexican students on Nov. 11, Mexico's envoy to the U.N., Adolfo Aguilar Zinser, characterized NAFTA as "a weekend fling." The U.S., he said, "isn't interested in a relationship of equals with Mexico, but rather in a relationship of convenience and subordination." While Zinser's remarks cost him his job, his words struck a chord. In an October survey by a leading pollster, only 45% of Mexicans said NAFTA had benefited their economy. That's down from the 68% who in November, 1993, saw the pact as a strong plus. With the U.S. in a slump for the past three years, Mexicans are experiencing the downside of their close commercial ties with the colossus. Mexico's economy will grow by 1.5% this year, a poor showing for a developing country.

Mexico believed NAFTA would make it the U.S.'s top workshop. China got the job

In a larger sense, Mexicans feel shortchanged by globalization. They thought they would be America's biggest workshop. That honor now belongs to China, which this year surpassed Mexico as the U.S.'s No. 2 supplier. Mexican policymakers signed trade agreements with a total of 32 countries, and as a result consumers got cheaper and better goods. Yet local manufacturers of everything from toys to shoes, as well as farmers of

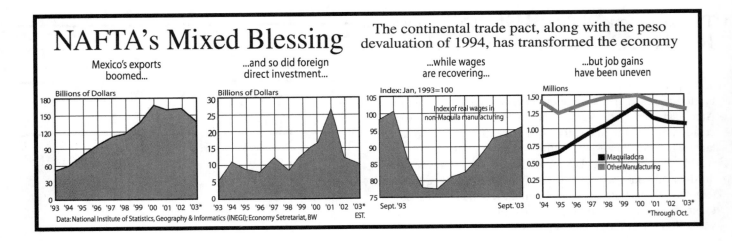

NAFTA's Mixed Blessing The continental trade pact, along with the peso devaluation of 1994, has transformed the economy

Mexico's exports boomed... ...and so did foreign direct investment... ...while wages are recovering... ...but job gains have been uneven

Data: National Institute of Statistics, Geography & Informatics (INEGI); Economy Secretariat, BW

rice and corn, struggle to survive the onslaught of cheap imports. Mexicans hoped NAFTA would generate enough jobs to keep them at home. Instead, the jobless flock in ever-greater numbers across the border. Reforms that pressed on Mexico before NAFTA—modernizing the electricity sector, overhauling the tax code, shoring up the crumbling schools—are an even more difficult sell now that power is split among several parties.

OPPORTUNITY KNOCKED

DO MEXICO'S woes disprove the value of free trade? Few would argue that NAFTA was a waste. "If we didn't have NAFTA, we'd be in far worse shape than we are today," says Andrés Rozental, president of the Mexican Council on Foreign Relations. If NAFTA has disappointed, it is in large part because the Mexican government has failed to capitalize on the immense opportunities it offered. "Trade doesn't educate people. It doesn't provide immunizations or health care," says Carla A. Hills, the chief U.S. negotiator in the NAFTA talks. "What it does is generate wealth so government can allocate the gains to things that are necessary." If a government doesn't allocate new wealth correctly, the advantages of free trade quickly erode. That is Mexico's plight. "NAFTA wasn't an end unto itself, but a means to something, and that something was precisely the need to go further in reform," says former Mexican President Carlos Salinas, one of NAFTA's principal architects. "It's like Alice in Wonderland—you have to run faster and faster if you want to stay in the same place. Globalization won't wait for you."

The outcome of Mexico's struggle to regain the momentum is of vital interest not just to Latin America but also to the U.S. The Bush Administration has made trade a vital part of its hemispheric agenda. Besides, the U.S. needs a stable, prosperous Mexico on its border to stem the flood of illegal immigration and drugs. Mexico's ability to get to the next stage will also show whether low-wage economies around the globe can hold their own against China. "Mexico cannot compete sewing brassieres and tennis shoes," says Roger Noriega, U.S. Under Secretary of State for the Western Hemisphere. "they cannot compete with China—who can? Mexico has to modernize so it can move forward."

NAFTA has already proven a powerful impetus to reform. Mexico did not hike its import tariffs when the peso crisis of 1994 hit. Encouraged, Washington stepped in with a $40 billion bailout package that helped Mexico stabilize its finances and return to the capital markets in just seven months. Although wrenching, the devaluation turbocharged NAFTA by dramatically lowering the costs of Mexican labor and exports. The government's fiscal discipline has earned the country a coveted investment-grade rating on its debt. And the current recession is mild by historic standards. Most analysts see growth quickening to 3.5% next year.

Yet even with a rebounding economy, Mexico will not generate enough jobs to accommodate its fast-growing

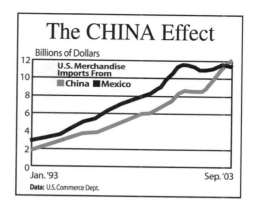

workforce. While U.S. companies praise the work of their Mexican employees, they now make it abundantly clear that there are other, cheaper locales. An assembly line worker in Mexico earns $1.47 an hour; his counterpart in China makes 59¢ an hour, according to a new report by McKinsey & Co. Top Delphi executives have warned for months that some production may be shifted to China because of the many cost advantages it offers. "Delphi and other automotive suppliers are courted every day by other countries, not only with lower-cost labor but also with new incentives and tax breaks," says David B. Wohleen, president for electrical, electronics, safety & interior. "Mexico will need to significantly pick up the pace to remain a competitive alternative," he warns.

No one feels the China threat more keenly than Daniel Romero, president of the National Council of the *Maquiladora* Export Industry. Mexico's *maquiladora,* which assembles goods for export using imported parts and components, had been around since the mid-1960s. Under NAFTA, the number of plants rose 67%, to 3,655 in seven years. Yet more than 850 factories have shut down since 2000, with many shifting to cheaper locales. Employment is down more than 20% from its peak of 1.3 million workers. Romero and a group of *maquiladora* managers traveled to China last year. They came away dispirited. "They have aggressive tax incentives, low salaries, very aggressive worker training, and a supply chain that allows them to have immediate access to the latest technology," says Romero.

The agriculture sector is suffering even more than the *maquiladoras*, as subsidized U.S. food imports flood the country. Some 1.3 million farm jobs have disappeared since 1993, according to a new report by the Carnegie Endowment for International Peace, a Washington think tank. "NAFTA has been a disaster for us," says Julián Aguilera, a pig farmer from Sonora. He and his peers have staged big demonstrations to protest a 726% increase in U.S. pork imports since the pact took effect. "Mexico was never prepared for this."

Nor was the U.S. As the *campesinos* lost their livelihood, they headed to the border. By most estimates, the number of Mexicans working illegally in the U.S. more than doubled, to 4.8 million between 1990 and 2000. Despite tightened security after September 11, hundreds of thousands continue to cross the border. The money sent back to their families will total $14 billion this year, more than the $10 billion Mexico expects in foreign direct investment.

The exodus has turned rural hamlets into ghost towns. Panindícuaro in Michoacán, one of Mexico's poorest states, has one the highest incidences of migration, with one out every seven people leaving. Panindícuaro's priest, Melesio Farías, recently held a funeral mass for a father in his mid-thirties who died trying to cross the Arizona desert. "I tell them to forget the U.S. and to work at home," says Farías. "But if Mexico can't offer them jobs, why should they?"

Salinas' band of technocrats and their successors didn't do enough to prepare vulnerable sectors for NAFTA'S onslaught. Long -promised programs to help 20 million *campesinos* switch to export crops never materialized. Nor has the government offered inducements to channel foreign investment into areas of the country where it is most needed. The six border states, along with the capital, nabbed 85% of foreign outlays year. Little has been done to foster a local supplier network for the import-dependent *maquiladoras*. Less than 3% of the industry's parts are sourced in Mexico. "Society at large and a good chunk of the economy have failed or refused to adjust to globalization," argues Luis Rubio, who heads the Center of Research for Development, a Mexico City think tank. "And the Mexican government has done absolutely nothing to help."

This laissez-faire attitude is in stark contrast to China. There, state-owned banks have bankrolled lavish investments in industrial parks, power plants, highways, and other infrastructure to provide low-cost facilities for foreign manufacturers. These multinationals had to source as many components as possible from domestic suppliers, and the government wasn't bashful about demanding transfers of technology to Chinese partners. Also, Beijing sealed off weak sectors like financial services or retailing. As a condition for entry into the WTO in 2001, China is phasing out these policies, but its domestic companies now have a head start.

Even if China-style tactics are not possible, Mexico could still hone its competitiveness. The PRI under Salinas took advantage of its monopoly on power to ram through painful reforms that paved the way for NAFTA. Now under a multiparty

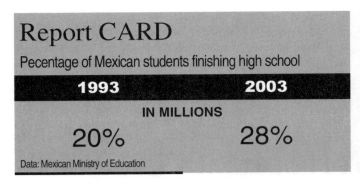

Report CARD

Pecentage of Mexican students finishing high school

1993	2003
IN MILLIONS	
20%	28%

Data: Mexican Ministry of Education

system, the politicians struggle to make difficult choices. Mexico will need to spend $50 billion to upgrade its power grid. But legislation to open the constitutionally protected sector to private investment has run aground on nationalist sentiment and union opposition, even while electricity rates are as much as 40% higher than in China. Grupo México, the world's third-largest copper producer, is considering moving refining operations to Amarillo, Tex., where electricity costs 4¢ per kilowatt-hour, vs. 8.5¢ in Sonora.

Education is another critical area where reform has stalled. William Spurr, head of the North American transport division of Canada's Bombardier, which builds railcars in Hidalgo, sees a need for more skilled workers. "There's a very good talent pool, but there aren't enough of them," he says."If I opened a plant in India, I'd have all the engineers and technicians I need." To be fair, the government's finances have been sapped by a $100 billion bank bailout after the peso crisis. Even under those circumstances, the number of science and engineering college grads has nearly doubled over the past decade, to 73,300. Yet that number still pales next to India, which graduated 314,000 students in those subjects, while China handed out diplomas to 363,000. Congress has so far foiled Fox's efforts to raise taxes to pay for improved education.

To get a glimpse of what the right training can do, consider the case of Tecnomec Agrícola, a maker of farm and earth-moving equipment based in Aguascalientes, in central Mexico. "We never had a tradition of exporting. NAFTA definitely changed that," says founder José Leoncio Valdés. It was hard going at first. "We couldn't get in to see people in the U.S. because we were from Mexico and they figured we were unreliable," recalls the 55-year-old engineer. Then in 2000, Valdés dispatched his son José to earn a degree in engineering and business administration at Massachusetts Institute of Technology. On his first spring break, young José conducted a weeklong session with Tecnomec managers. He used Lego blocks to build a replica of the factory and figure out how to better track inventory, boost quality, and control waste. Tecnomec soon boosted productivity by 21%. Now its exports total $1.4 million a year, nearly a quarter of the company's annual sales.

Mexico could use more Tecnomecs. Just 50 companies account for half of all of Mexico's exports—and the top tier is dominated by multinationals. Thousands of other Mexican businesses have gone under in the face of competition."We are at a watershed," says Jaime Serra Puche, Mexico's chief NAFTA negotiator. "Either we take the steps to become a true North American country or we just become a big Central American country."

Serra Puche is one of many prominent Mexicans trying to figure out how to improve NAFTA. "If we were going to do it all over again today, I would insist on introducing a lot of considerations," says President Fox, who believes that NAFTA should be modeled more on the European Union, with provisions for free movement of labor and cross-border grants to compensate poorer countries for the dislocations. Proposals for a single currency, a North American energy cooperation plan have also surfaced. But don't expect any breakthroughs soon—not while the U.S. heads into elections and trade has re-emerged as a contentious issue.

So for now the burden will remain on Mexico. Salvador Kalifa, an independent economist based in Monterrey, recalls that when Spanish conqueror Hernán Cortés reached Mexico, he burned his boats to prevent crew members from fleeing. "With NAFTA, we burned our boats and threw ourselves into globalization," says Kalifa. "There is no turning back."

Index

Index

water-insurance system, 135
water-sharing arrangements, 135
Wilford, John Noble, 2
Wilhite, Donald, 133
Wisconsin, use of coal by power
	companies in, 47, 48, 49
World Food Program (WFP), 130, 131

X

Xcel Energy, 46–47
xeriscape, 134

Z

Zambia, AIDS in, 115
Zhu Rongji, 42–43

Test Your Knowledge Form

We encourage you to photocopy and use this page as a tool to assess how the articles in *Annual Editions* expand on the information in your textbook. By reflecting on the articles you will gain enhanced text information. You can also access this useful form on a product's book support Web site at *http://www.mhcls.com/online/*.

NAME:

DATE:

TITLE AND NUMBER OF ARTICLE:

BRIEFLY STATE THE MAIN IDEA OF THIS ARTICLE:

LIST THREE IMPORTANT FACTS THAT THE AUTHOR USES TO SUPPORT THE MAIN IDEA:

WHAT INFORMATION OR IDEAS DISCUSSED IN THIS ARTICLE ARE ALSO DISCUSSED IN YOUR TEXTBOOK OR OTHER READINGS THAT YOU HAVE DONE? LIST THE TEXTBOOK CHAPTERS AND PAGE NUMBERS:

LIST ANY EXAMPLES OF BIAS OR FAULTY REASONING THAT YOU FOUND IN THE ARTICLE:

LIST ANY NEW TERMS/CONCEPTS THAT WERE DISCUSSED IN THE ARTICLE, AND WRITE A SHORT DEFINITION:

We Want Your Advice

ANNUAL EDITIONS revisions depend on two major opinion sources: one is our Advisory Board, listed in the front of this volume, which works with us in scanning the thousands of articles published in the public press each year; the other is you—the person actually using the book. Please help us and the users of the next edition by completing the prepaid article rating form on this page and returning it to us. Thank you for your help!

ANNUAL EDITIONS: Geography 06/07

ARTICLE RATING FORM

Here is an opportunity for you to have direct input into the next revision of this volume.
We would like you to rate each of the articles listed below, using the following scale:

1. **Excellent: should definitely be retained**
2. **Above average: should probably be retained**
3. **Below average: should probably be deleted**
4. **Poor: should definitely be deleted**

Your ratings will play a vital part in the next revision.
Please mail this prepaid form to us as soon as possible.
Thanks for your help!

RATING	ARTICLE
_____	1. The Big Questions in Geography
_____	2. Rediscovering the Importance of Geography
_____	3. The Four Traditions of Geography
_____	4. The Power of Place
_____	5. The Changing Landscape of Fear
_____	6. Watching Over the World's Oceans
_____	7. After Apartheid
_____	8. The Race to Save a Rainforest
_____	9. Global Warming
_____	10. Environmental Enemy No. 1
_____	11. A Great Wall of Waste
_____	12. The New Coal Rush
_____	13. The Rise of India
_____	14. Between the Mountains
_____	15. A Dragon with Core Values
_____	16. Where Business Meets Geopolitics
_____	17. Oil Over Troubled Water
_____	18. Central Washington's Emerging Hispanic Landscape
_____	19. Drying Up
_____	20. Living with the Desert
_____	21. Deep Blue Thoughts
_____	22. An Inner-City Renaissance
_____	23. Mapping Opportunities
_____	24. Geospatial Asset Management Solutions
_____	25. Internet GIS: Power to the People!
_____	26. The Future of Imagery and GIS
_____	27. Calling All Nations
_____	28. Mapping the Nature of Diversity
_____	29. Fortune Teller
_____	30. Asphalt and the Jungle
_____	31. A City of 2 Million Without a Map
_____	32. AIDS Infects Education System in Africa
_____	33. The Longest Journey
_____	34. China's Secret Plague
_____	35. Farms Destroyed, Stricken Sudan Faces Food Crisis
_____	36. Dry Spell
_____	37. Turning Oceans Into Tap Water
_____	38. Putting the World to Rights

RATING	ARTICLE
_____	39. Mexico: Was NAFTA Worth It?

(Continued on next page)

ANNUAL EDITIONS: GEOGRAPHY 06/07

BUSINESS REPLY MAIL
FIRST CLASS MAIL PERMIT NO. 551 DUBUQUE IA

POSTAGE WILL BE PAID BY ADDRESEE

McGraw-Hill Contemporary Learning Series
2460 KERPER BLVD
DUBUQUE, IA 52001-9902

NO POSTAGE
NECESSARY
IF MAILED
IN THE
UNITED STATES

ABOUT YOU

Name _____   Date _____

Are you a teacher? ❑   A student? ❑
Your school's name _____

Department _____

Address _____ City _____ State ____ Zip ____

School telephone # _____

YOUR COMMENTS ARE IMPORTANT TO US!

Please fill in the following information:
For which course did you use this book? _____

Did you use a text with this ANNUAL EDITION? ❑ yes ❑ no
What was the title of the text? _____

What are your general reactions to the *Annual Editions* concept? _____

Have you read any pertinent articles recently that you think should be included in the next edition? Explain. _____

Are there any articles that you feel should be replaced in the next edition? Why? _____

Are there any World Wide Web sites that you feel should be included in the next edition? Please annotate. _____

May we contact you for editorial input? ❑ yes ❑ no
May we quote your comments? ❑ yes ❑ no